'The potato is perhaps the most unfairly maligned food!'

Sheila Bingham's book contains many such surprising statements, including the facts about white bread versus the preferred wholemeal. This book is based on her sound scientific knowledge as a dietitian, and as such contains a wealth of new and little known facts.

Technology has altered many of the foods we eat, and not always for the better! Her book leads the reader through the pitfalls of convenience foods, manufacturers' 'small print', and diet fads and fallacies, to a sound understanding of nutrition and the road to better health.

Nutrition: A Consumer's Guide to Good Eating

Originally Published as
Dictionary of Nutrition

Sheila Bingham

CORGI BOOKS
A DIVISION OF TRANSWORLD PUBLISHERS LTD

Nutrition: A Consumer's Guide to Good Eating
A CORGI BOOK 0 552 10664 X

Originally published in Great Britain by
Barrie & Jenkins Ltd, under the name of
DICTIONARY OF NUTRITION

PRINTING HISTORY

Barrie & Jenkins edition published 1977
Corgi edition published 1978

This book is set in Times 10/11 pt.

Corgi Books are published by Transworld Publishers, Ltd.,
Century House, 61–63 Uxbridge Road, Ealing,
London, W5 5SA

Made and Printed in Great Britain by
Richard Clay (The Chaucer Press), Ltd., Bungay, Suffolk.

Nutrition: A Consumer's Guide
to Good Eating

INTRODUCTION

Diet is popularly thought of as merely an unpleasant way of getting slim: a panic response to fat. Books on slimming suggest that food is the enemy, meals the battlefield. They teach defensive measures, but little insight.

In reality, of course, diet means your daily food, and diets can be good or bad. They can keep you slim, or make you fat. This book is about positively using food as an ally, so that panic measures become unnecessary. But just as there are techniques to learn before driving a car, so there are basic facts to find out about food before eating it. To know how to use food intelligently, to get the fullest value from it, it must be understood. Food may appear commonplace, but the range of foods, the many **nutrients** each different food supplies, the **additives** used in processing, and the way each food is cooked, all add to the complexities of eating. It is no longer safe to assume that good taste is a reliable guide to goodness, nor to rely on habits passed from one generation to the next. Our way of life has been changed by technology, and food has changed with it.

This Guide will give you, the consumer, a better understanding of nutrition, the science of food in relation to health. It is neither a get-slim-quick nostrum, nor an over-complicated scientific text, but a straightforward guide to healthy eating. It is not about recipes, but it is a companion to use with cooking. It will help you decide what types of food you need in your daily diet; how to find the best nutritional value; how to retain essential nutrients in storage and cooking; and how to avoid food poisoning. After reading it, you will know whether or not a packet of instant **potatoes** contains **vitamin C**; if **soya** meat is as good as real **meat**; if **monosodium glutamate** will do you any harm; and whether you should buy wholemeal **bread**. You will also know what is nutritious about **nuts**; what is in your daily pinta; whether **honey** is for the bees; and what is good about going to work on an **egg**. The established facts about diet and disease are explained, and popular fallacies examined. The Guide tells you what all the essential nutrients in your food actually do; how much you need; and if and when you need to take extra. It does not claim to reveal any miraculous cures for baldness, wrinkles or irritability, but it does, for example, help you decide whether you need a low **cholesterol** diet; how to choose a **slimming diet**; and what is meant by a **bland diet**. It explains why children need **breakfast**; why breast **milk** is better for babies; what to eat for a **balanced diet** in pregnancy; and the facts behind recent controversies like **fibre, fat** and **fluoride**.

To indicate that they are explained in more detail under their own entry, foods, nutrients, additives and other terms are sometimes printed in **bold type**, as in this introduction. A list of the 200 entries is shown on page 13. At the end of each food entry is a table of nutrient contents for individual foods. For instance, to find out the nutrients in beef, look under **meat**; in an orange, look under **fruit**. The introductions to most entries contain the salient points: those who would like to find out more should read on.

For those not entirely used to metric units, like kilograms and litres, approximate measures in imperial units follow in brackets. Most people may not be aware that the familiar **Calorie**, which effectively means **energy** in food, has already been replaced by the more general energy unit, the **joule**, in scientific texts and journals. However, to lessen an inevitable source of confusion to those unused to scientific terms, Calories (or kilocalories) have been retained in the text. For those who wish to practise before the inevitable general changeover, kilocalories can be converted to kilojoules by multiplying by 4.2. Both units have been used in the energy section and in the food tables.

ACKNOWLEDGEMENTS

Jill Cave Browne Cave, BSc, SRD, Chief Dietitian at University College Hospital, London, and Jane Griffin, BSc, SRD, made many helpful suggestions and spent much time checking the final draft of this book. The advice of Alison Paul, BSc, at the Dunn Nutritional Unit, Cambridge, on the recent changes in the composition of foods, was also invaluable, and Dr W. P. T. James, Assistant Director of the Dunn Nutritional Unit, was no less kind in discussing particular problems. To all of these people I offer my sincere thanks; should there be any inaccuracies in the text, the fault lies entirely with the author.

I am indebted to Mr Tither and his staff for permission to use the library at the Ministry of Agriculture, Fisheries and Food, and for their help in locating original articles; and to the Controller, Her Majesty's Stationery Office for permission to base the figures for the nutrient content of food on those published in *The Composition of Foods* by R. A. McCance and E. M. Widdowson, and to reproduce the tables on pages 41, 142, 378 and 379.

I would also like to thank John Knowler—whose suggestion it was that I write the book, based on his original idea—for his constructive advice and criticism of the first drafts; Raymond Hawkey for his help with particular aspects; Dr and Mrs A. M. Joekes, who made it possible for part of the book to be written in the idyllic surroundings of the Villa Pipistrello; and in particular Roger Bingham for his discerning editing of the earlier drafts and for his unfailing moral and financial support, without which the book would never have been completed.

LIST OF ENTRIES

Entry	Page	Entry	Page
Acetic acid	17	Bread	47
Additives	17	Breakfast	56
Agar	21	Butter	57
Alcohol	22		
Alginates	25	Caffeine	58
Allergy	26	Calcium	58
Amino acids	27	Calorie	62
Anaemia	32	Carbohydrate	63
Anorexia nervosa	32	Carotene	65
Antibiotics	33	Carragheenin	68
Antioxidants	33	Cellulose	68
Arginine	34	Cereals	68
Arrowroot	35	Cheese	71
Ascorbic acid	35	Chlorine	73
Ash	36	Chocolate and	
Atherosclerosis	36	confectionery	74
		Cholesterol	75
Balanced diet	37	Choline	78
Barley	41	Chromium	78
Beer	42	Citric acid	80
Beriberi	44	Cobalt	80
Beverages	44	Cocoa	80
Bioflavonoids	45	Coeliac disease	81
Biotin	45	Coffee	83
Bland diet	46	Colour	83
Bran	46	Convenience foods	86

Entry	Page	Entry	Page
Cooking fats	87	Gluten	158
Copper	88	Glycerol	159
Cream	90	Glycine	159
Cysteine	92	Glycogen	160
		Gout	160
Dextrins	92	Gravy	163
Diabetes	92		
Diet	96	Health foods	163
Digestion	97	Heart disease	165
Disaccharides	99	Heat in foods	168
Dyspepsia	99	Herbs	168
		Histidine	169
Edible gums	101	Honey	169
Eggs	101	Hydrolysed protein	170
Electrolytes	106		
Emulsifiers and		Ice cream	170
stabilisers	107	Inborn Errors of	
Energy	109	Metabolism	171
Enzymes	114	Inositol	172
Extracts	116	Inulin	173
		Invert sugar	173
Fat	117	Iodine	173
Fatty acids	120	Iron	176
Fibre	122	Isoleucine	180
Fish	128		
Flavour	133	Joule	180
Flour	136		
Fluorine	139	Kwashiorkor	181
Folic acid	146		
Fructose	149	Labelling of food	182
Fruit	150	Lactic acid	184
		Lactose	184
Galactose	155	Lecithin	185
Gelatine and Jelly	156	Leucine	186
Glucose	157	Linoleic acid	186
Glutamic acid	158	Lipid	188

Entry	Page	Entry	Page
Lysine	190	Pantothenic acid	245
		Para amino benzoic acid	246
Magnesium	191	Pectin	247
Maize	194	Pellagra	247
Malabsorption	196	Phenylalanine	248
Malnutrition	197	Phospholipids	249
Maltose	197	Phosphorus	250
Manganese	198	Phytic acid	251
Marasmus	198	Poisons in food	252
Margarine	199	Polyphosphates	267
Meat, poultry and offal	201	Polysaccharides	267
Metabolism	209	Potassium	267
Methionine	210	Potatoes	268
Milk	211	Preservatives	271
Millet	218	Preserves	274
Mineral hydrocarbons	219	Protein	275
Mineral water	220	Pulses	281
Minerals	220	Pyridoxine	285
Molybdenum	220		
Monosaccharides	221		
Monosodium glutamate	221	Raffinose	288
		Raising agents	288
Nickel	222	Recommended intake	289
Nicotinic acid	223	Riboflavin	290
Nitrite and nitrate	226	Rice	292
Nitrogen	228	Rickets	294
Novel protein	228	Roughage	295
Nutrients	230	Rye	295
Nuts	231		
		Sago	296
Oats	235	Salt	297
Obesity	236	Scurvy	297
Oils	240	Selenium	298
Osteomalacia	243	Silicon	299
Osteoporosis	244	Slimming diets	300
Oxalic acid	244	Sodium	307

LIST OF ENTRIES

Entry	Page	Entry	Page
Soft drinks	310	Tryptophan	336
Solvents	311	Tyrosine	336
Sorbitol	312		
Soups	312	Ulcers	337
Soya beans	314		
Spices	314	Valine	338
Spirits and liqueurs	315	Vanadium	338
Starch	316	Vegetables	338
Starvation	318	Vegetarianism	343
Sucrose	319	Vinegar	344
Sugar	321	Vitamins	344
Sulphur	321	Vitamin A	345
Sweeteners	322	Vitamin B complex	349
		Vitamin B_{12}	350
		Vitamin C	353
Tapioca	323	Vitamin D	358
Tartaric acid	324	Vitamin E	361
Tea	324	Vitamin K	364
Teeth	325		
Textured vegetable		Water	366
protein	329	Wheat	368
Therapeutic diets	329	Wine and cider	368
Thiamin	330		
Threonine	333	Yeast	370
Tin	334	Yoghurt	373
Trace elements	334		
Triglycerides	335	Zinc	374

ACETIC ACID

Main acid in **vinegar**. Occurs naturally in small amounts in the body and most foods. Synthetic acetic acid is added to pickles, sauces and fruit-flavoured foods as a preservative and acidifier. Calcium, sodium and potassium salts (. . . acetate) are used as buffers (see Additives).

ADDITIVES

Non-nutritive substances added intentionally to food to increase shelf life, to assist in processing and to improve palatability.

Estimates of the number of food additives in use vary. About 250 substances are specifically permitted in food in Britain, including 'natural' substances like annatto, butter colour. There are a further 1500 flavours and 12 groups of additives in use. Foods and derivatives of foods—like modified **starch** and **hydrolysed protein**—although they perform the same functions as additives, are not subject to control. The main types of additives are listed in the table overleaf. Some individual additives have a dual role, for instance **glycerol** is a solvent and humectant (see below), **polyphosphates** are sequestrants and buffers (see below).

The use of food additives is controlled by the Food and Drugs Act 1955, which—among other general provisions—authorises the Minister of Agriculture, Fisheries and Food and the Secretary of State for Social Services jointly to make regulations concerning food additive content, labelling and composition of food. The ministers are advised by the Food

Additives and Contaminants Committee and the Food Standards Committee. Codes of Practice used by the food industry provide guidance in legal proceedings, but do not carry the force of law.

Until comparatively recently, only food additives known to be harmful were specifically banned from food. The Food and Drugs Act also made it an offence to sell food unfit for human consumption and not of the nature, substance and quality demanded, or injurious to health. This situation has been improved by the introduction of permitted lists: eventually only additives specified in government regulations will legally be added to food. There are 10 permitted lists—see table; lists for 13 other groups (including flavours) are being planned. Permitted additives must be indispensable, they must not deceive the customer as to the nature and quality of food, and they should not be used in greater quantities than necessary. Some—particularly the preservatives—are only allowed up to certain maximum quantities and/or in specified foods.

Main types of food additives

Covered by permitted lists
  Preservatives
  Colours
  Antioxidants   —see individual
  Emulsifiers and stabilisers   entries
  Sweeteners
  Solvents
  Mineral hydrocarbons
Bleaches and improvers   —see flour
Miscellaneous   —see page 20
  Acids and buffers
  Bases
  Humectants
  Sequestrants
  Propellants
  Glazing agents

Antifoaming agents
Anticaking agents
Release agents
Firming and crisping agents
Liquid freezant
Packaging gas
Not yet covered by permitted lists
Flavours   —see flavours
Artificial smoke solutions
Bulking agents
Clouding agents
Crystallisation inhibitors
Dilutents
Encapsulating agents
Enzymes   —see page 116
Excipients

Flavour modifiers   } —see monosodium
                                    glutamate

Mordants
Tableting aids
Whipping aids

Permitted food additives are thought to be safe for both
long- and short-term use. On the available evidence they are
not harmful to health, growth, reproduction and life span. In
some respects a permitted additive is safer than many
'natural' products because it is a small amount of one known
chemical rather than a mixture of a great number of chemicals—some of which are unknown—as in food and natural
products. However, safety tests are necessarily restricted to
animals; apart from the differences in the way man and
animals utilise food and additives, interaction of foods and
additives (causing toxic compounds) cannot always be predicted. Consequently the effect of an additive on human
health is uncertain until it has been in use for many years.
New evidence, perhaps after the additive has been tested on a
different animal, may raise doubts and it may be taken off
the permitted list.

Animal feeding trials determine whether a new additive is toxic (some are not, even in massive doses) and the maximum amount that has no effect in the diet. Additives that have no 'no-effect level' (being toxic at any dosage) or are carcinogens are not permitted. The 'no-effect level' is then usually arbitrarily divided by a safety factor of 100 and called the Acceptable Daily Intake. When the quantity that would be eaten in a day's diet is greater than the Acceptable Daily Intake, the additive is restricted (to avoid overconsumption) to those foods only eaten occasionally and/or a maximum permissible amount is specified.

Many new additives were developed for use in **convenience foods**. These—particularly the 'instant' type—tend to contain more additives than staple or lightly processed foods. See also Labels on food. Most fresh foods contain no additives but apples and pears may have anti-oxidant, and citrus fruit mineral hydrocarbon on the skin. Wholemeal flour (but not wholemeal bread) is not permitted to contain any additives.

Miscellaneous additives

The Miscellaneous Additives in Food Regulations 1974 permit approximately 130 substances, some of which are allowed in specific amounts only in certain foods.

Acids and buffers—added to give an acid taste (acidifiers). They include acetic, citric, lactic, malic, phosphoric, hydrochloric, sulphuric, nicotinic acids and acid sodium aluminium phosphate (in baking powder). Buffers control acidity and are usually sodium, calcium and potassium salts of weak acids—for example, calcium lactate.

Bases (alkalies) also control acidity. They include calcium, potassium, magnesium and sodium hydroxides and carbonates.

Humectants—keep foods like cakes soft, by preventing the food from drying out. **Sorbitol** and **glycerol** and sodium and potassium lactate are examples of humectants.

Sequestrants—inactivate stray minerals, thus preventing

deterioration from rancidity (see Antioxidants). Include **glycine** and, in certain substances (including canned fish to prevent the formation of glass-like struvite), sodium calcium edetate. Citric acid and polyphosphates are also sequestrants.

Propellants—used in aerosols (like whipped cream). The gases nitrogen, carbon dioxide and nitrous oxide are permitted.

Glazing agents—coated over food to give a shiny surface, which helps to preserve it. Include beeswax, gelatin and shellac (mineral hydrocarbons are also glazing agents used in sweets).

Antifoaming agents—needed to prevent foam in foods like evaporated milk and fruit juice during canning. Include dimethylpolysiloxane (see silicone).

Anticaking agents—prevent powders from clumping together and are added to salt and most dry powder foods, like cake mixes and coffee whiteners. Include silicates (see silicone).

Release agents—prevents sticking of food to utensils. Includes sperm oil but glycerol and mineral hydrocarbons are also release agents.

Firming and crisping agents—Calcium salts added to canned vegetables to give 'bite'—see pectin. Includes calcium chloride.

Packaging gas and liquid freezant—used to exclude air from packs and to freeze foods. For example nitrogen.

AGAR

An extract of red seaweed, used as a thickener and stabiliser (see Emulsifiers) in many foods, for instance ice cream, soups, sweets. It dissolves in hot water and sets to a jelly when cooled: it is sometimes used with gelatine for meats in jelly.

Agar is a **polysaccharide**, composed of chains of **galactose**. Although it is not affected by the digestive **enzymes** and contributes to roughage (see fibre), it may have to be avoided in galactose free diets. Small amounts may be fermented by bacteria in the large bowel and the liberated galactose may pass into the blood stream.

ALCOHOL—chemical names ethanol, ethyl alcohol

Produced by yeasts growing on liquids containing sugars and starches. Other substances—from food used as a source of carbohydrate, from added constituents, and from those produced by the yeast during fermentation—contribute to the characteristics of different alcoholic beverages.

Alcoholic **beverages** are sources of **energy**. The alcohol supplies 7 kilo**calories** per gram and, in all except spirits, sugars supply 4 kilocalories per gram. Liqueurs are the most calorific. Beers and wines also contain B **vitamins** and **minerals**. Alcoholic drinks produced by fermentation alone (beers and wines) do not contain more than 13 grams of alcohol per 100 mls: above this level the yeast is inhibited. In spirits and liqueurs the alcohol is concentrated by distillation from the ferment. Fortified wines are intermediate in alcohol content between wines and spirits. The table overleaf shows approximate alcohol and energy contents of average measures of alcoholic drinks. See individual entries for more detailed information.

Alcohol contents are expressed as percentage by weight (usually taken as grams per 100 mls of drink), percentage by volume (mls per 100 mls of drink) or degrees proof. Percentage by volume can be approximately converted to percentage by weight by multiplying by 0·8.

The original way of measuring proof was to mix a sample of alcohol with a small quantity of gunpowder and set light to it. 'Proof' spirit contained sufficient alcohol to ignite the gunpowder: in an underproof sample the alcohol would burn but the relative excess of water made the gunpowder too damp to explode. The modern definition of proof spirit (in the U.K.) is that it contains approximately 50% of alcohol by weight and 57% by volume. For instance, 70° proof gin contains 70% of proof spirit by volume—i.e. $70/100 \times 57 = 40$ mls of alcohol per 100 mls of gin (or approximately $40 \times 0.8 = 32$ grams of alcohol per 100 mls of gin).

APPROXIMATE ALCOHOL, ENERGY AND CARBOHYDRATE CONTENTS OF
ALCOHOLIC DRINKS

Drink	Average measure	Alcohol grams	Carbohydrate grams	Energy kilojoules	kilocalories
Beer and cider	300 mls ($\frac{1}{2}$ pint)	7–32	5–22	300–1260	70–300
Table wine	150 mls (1 glass)	13–15	0·3–9	380–590	90–140
Spirits, 70° proof	1/6 gill	8	0	230	55

About one-fifth of the alcohol in a drink is absorbed through the walls of the stomach into the blood stream. Effervescent drinks and spirits are absorbed more quickly, but food or—in some people—fat (for example in milk) taken in advance delay absorption from the stomach. The rest of the alcohol is steadily absorbed into the blood stream from the small intestine (see Digestion) at a constant rate unaffected by food. Most people will have absorbed all the alcohol in a drink after about 2 hours.

Once in the blood stream, alcohol diffuses into the water in all cells of the body. Nerve cells are particularly susceptible to its inhibitory effects. Areas of the brain responsible for self-control and criticism are depressed first, promoting euphoria and lack of judgement. Later, taste, smell, co-ordination (reaction time), hearing and vision are adversely affected. Symptoms of intoxication are apparent—especially to those unused to alcohol—when the blood contains about 100 milligrams of alcohol in 100 mls. This is equivalent to 50 grams of alcohol (contained in about 3 pints of bitter beer) for an 11 stone (70 kilogram) man. In Britain it is an offence to drive a car when the blood alcohol is greater than 80 milligrams per 100 mls. This is roughly 40 grams of alcohol (contained in about $2\frac{1}{2}$ pints of bitter beer) for an 11 stone man, but less for those with a smaller build.

The liver is chiefly responsible for ridding the body of alcohol, converting it to carbon dioxide and water and releasing energy in the process. Most of the energy released is converted to **fat** which is used immediately or stored for later use. A small quantity (5 to 10% of the intake) is passed out of the body unchanged in urine, breath and sweat.

The rate at which alcohol is used by the body is constant for each individual: some people have half the average rate, others double. On average an 11 stone man would need $1\frac{1}{2}$ hours to use up 10 grams of alcohol (in just over $\frac{1}{2}$ pint of bitter beer). Women have a slower rate and are generally smaller than men, so that the intoxicant effects for the same dose of alcohol are apparent for longer.

It is possible to speed up the rate of alcohol eliminated by the liver by eating **fructose**, but exercise, coffee and vitamins have no effect. The stimulatory effect of fructose seems to vary between individuals—30 to 90 grams may be required. Regular use of alcohol and fructose is inadvisable because more fats are produced and **gout** may be exacerbated. See also lipid. Hangovers are mainly due to the non-alcoholic—congener—constituents of alcoholic drinks. Vodka and gin contain the least congeners, brandy and red wine the most. Additionally alcohol increases the flow of urine, leading to dehydration: hangovers can be moderated by drinking plenty of water, preferably before sleep.

In moderation (up to 50 grams for most people), alcohol is not harmful and may be of benefit in some circumstances. By relieving tension it is conducive to a good appetite and sleep. It may also calm intestinal colic and dilate the blood vessels of the skin, promoting a feeling of warmth. (Asians seem to be more susceptible to flushes than Europeans.) However, the warming effects of alcohol are superficial: when blood is diverted to the skin, the body actually loses heat. For this reason it should not be given to persons suffering from exposure. Alcohol is also inadvisable after severe exercise or lack of food because it suppresses the regeneration of **glycogen**, necessary for the maintenance of an acceptable blood **glucose** level. The resultant fall in the blood sugar level has sometimes induced coma and further lowering of the body temperature, with occasional fatal results. Children and diabetics treated by drugs or insulin are more susceptible to the blood sugar lowering effect of a moderate to large dose of alcohol.

Consumption of alcohol in Britain has increased markedly over recent years—by 80% for wine, 25% for beer and 30% for spirits between 1962 and 1972. Partly because of its availability, it is the most common drug of addiction. Chronic alcoholism affects at least 1% of adults in England and Wales and 2% in Scotland. At least a further 3–4% of adults are heavy or problem drinkers.

When regularly taken in excess (over ½ bottle of spirits per day) alcohol depresses appetite, irritates the stomach (exacerbating ulcers) and affects the utilisation of many nutrients. Consequently chronic alcoholics, and some heavy drinkers, are likely to suffer from **malnutrition**. There is evidence that alcohol is also primarily responsible (rather than malnutrition) for liver damage. Initially fat accumulates in liver cells but can be eliminated if the consumption of alcohol is reduced. Several years' excessive alcohol consumption is one cause of liver inflammation (hepatitis) which is usually followed by cirrhosis: a condition in which the liver cells die and are replaced by scars. An extensively cirrhosed liver cannot support life.

ALGINATES

Extracted from kelp seaweed, alginates are used in many foods as **emulsifiers**, thickeners, stabilisers, and gelling agents. For instance, instant puddings, ice cream, salad cream, processed cheese, artificial cream, and whipped cream may all contain alginates. Other uses include the manufacture of fruit 'pieces' for pie fillings from comminuted fruit, and artificial cherries.

Alginates are calcium, potassium, sodium, iron, magnesium or ammonium salts of alginic acid (in the same way that sodium chloride (salt) is the sodium salt of hydrochloric acid)—a **polysaccharide** which contributes to dietary **fibre**.

ALLERGY

About 10% of people inherit the tendency to allergy. It is an unusual sensitivity to one or more substances in food, or in the air (for instance pollen, the housedust mite), or in contact with the skin (for example fur, cosmetics). The foods most commonly incriminated are milk, eggs, wheat, fish (especially shellfish), bananas, tomatoes, strawberries, meat (particularly pork), nuts and chocolate. Allergies to milk, wheat and eggs are commonest in children: after the age of five, food allergies tend to be replaced with other allergies, for instance to pollen.

The substance causing the allergy is called an allergen: allergens are usually **proteins** or glycoproteins (complexes of protein and **carbohydrate**). Normally proteins in food do not cause allergies because they are completely split to their component amino acids during **digestion** before entering the blood stream. If, however, tiny quantities of protein or peptides are absorbed intact, special types of antibodies may be formed. The antibodies fix on to cells (particularly those called 'mast' cells) in the skin and mucous surfaces (linings of the respiratory system—lungs, nose, sinuses, etc.—and digestive system). When the allergen next enters the blood stream, the union of allergen and antibody causes the mast cells to rupture, releasing powerful substances, including histamine and serotonin. Individual symptoms of allergy depend on whether the antibodies are fixed in the skin, respiratory system or digestive system. Nettle rash, eczema, hay fever, asthma, stomach cramps, vomiting, or diarrhoea may result. However, suggestions that mental illnesses are allergic reactions are based on very dubious evidence.

Some people have only mild allergies—others recover spontaneously. When there is an immediate reaction it is easy to pinpoint and avoid the food concerned, but other allergens may only be established from detailed investigations at specialised clinics. Sometimes food and cosmetics may contain common allergens. For instance, a person may become sensitised to a substance similar to benzoic acid (a **preservative**) in sun lotions, and later react to benzoic acid in pickles, soft drinks and instant coffee. Once an allergy has been confirmed, doctors should be informed. Severe reaction, anaphylactic shock, can follow injection of the allergen—for instance people allergic to egg can be made very ill by immunising injections prepared with egg, such as some 'flu vaccines. Others, allergic to pork, are sensitive to insulins obtained from pigs.

AMINO ACIDS

The building blocks of **protein**. As a rough analogy, proteins are built up in living cells from 20 individual amino acids just as words are built up from individual letters in the alphabet. In the case of proteins, though, the 'word' is several hundreds of 'letters' long. The type of protein is determined by the sequence of amino acids: the protein in hair—keratin —is very different from a protein in milk—casein. The sequence is dictated by the genetic machinery (genes) in each cell and different species of animals and plants make different proteins.

Plants make amino acids from **nitrogen** salts in the soil. Like sugars and fats, amino acids contain carbon, hydrogen and oxygen, but in addition they all contain nitrogen (3 amino acids also contain **sulphur**). Animals cannot make all the amino acids they require and must obtain their supplies ultimately from plants. During **digestion**, proteins are split to their constituent amino acids, which are absorbed into the blood stream.

Every human cell contains protein, which is continually broken down. Additionally, cells die and have to be replaced. Amino acids in the correct proportion are withdrawn from the blood stream by cells throughout the day, and used to resynthesise new cell (body) protein to make good these losses. The potentially toxic nitrogen, released from the old amino acids, is eventually converted to urea—a relatively harmless substance later filtered out of the blood stream by the kidneys and excreted into the urine. The carbon, hydrogen and oxygen fraction is used to supply energy.

Eight of the 20 amino acids are not produced naturally in the human body. For resynthesis of new body protein to continue, these 'essential amino acids'—methionine, trypto-phan, threonine, valine, isoleucine, leucine, phenylalanine, and lysine (see individual entries)—must be provided from proteins in the daily diet. Histidine is called semi-essential: adults seem able to manage without it, but growing children cannot. Different quantities of each essential amino acid are needed—for instance adults probably need about $\frac{1}{4}$ gram of tryptophan and $\frac{1}{2}$ gram of lysine each day. The other amino acids—the non-essential amino acids—are present in food proteins and can also be made in the body, provided there is sufficient nitrogen available from the breakdown of un-wanted amino acids in food and discarded proteins.

Proteins in food supply different proportions of essential amino acids. Those which approach the ideal (well balanced) proportion when eaten alone are called 'good quality'. More poorer quality protein needs to be eaten to fulfil the daily requirement for essential amino acids than good quality protein. In general, animal proteins (old name: first class) like meat, milk, fish and eggs, are of better quality than vegetables like peas, wheat and maize, but there are notable exceptions. For instance soya protein approaches the value of cheese protein, some combinations of vegetable proteins are able to supplement each other (see below), and gelatine is very poor.

Good quality proteins are only used efficiently if there is sufficient energy in the diet. Theoretically, then, although the adult daily requirement for essential amino acids would be met by six eggs, if they were eaten alone the diet would only

contain 550 kilocalories. In such a low energy diet most of
the amino acids would be diverted away from the repair of
body protein and used to supply energy. At least another
1000 kilocalories would have to be eaten (either as 12 more
eggs or other food) to avoid eventual starvation and protein
deficiency.

The quality of proteins eaten is of less importance in
Britain and other Western societies. Moderate quality
protein-containing foods—like the cereals—are in abundant
supply. They contain some of all the essential amino acids
and—provided sufficient is eaten—the daily needs for essential amino acids (but **not** other essential nutrients) will be
met. The majority of Western diets also contain plenty of
animal proteins, which, because of their better balance of
essential amino acids, will make good any deficiency.

The quality of protein eaten is of greater importance when
either the diet is low in protein (from any source) or when it
is lacking in variety with animal protein foods in short
supply. These conditions most frequently occur in underdeveloped countries—though low protein diets are prescribed for some liver and kidney diseases. Some 'fad' diets
are also low or deficient in good quality protein.

Measurement of protein quality

For practical purposes, the quality of a protein is estimated
by comparing its content of essential amino acids with those
of 'reference protein'. Reference protein is a purely theoretical concept containing an ideal proportion of essential
amino acids assumed to fulfil daily needs when it is the sole
source of protein in the daily diet.

There is a dual effect when a protein contains less than the
assumed ideal proportion of an essential amino acid in
reference protein. If it is relatively lacking by, for instance,
half of one essential amino acid, only half of all the other
essential amino acids can be used to make new body protein.
The rest is discarded—or wasted—and used to supply
energy. Twice as much of this protein, compared to reference

protein, would therefore have to be eaten to fulfil the daily
needs for essential amino acids. The most deficient amino
acid—the weakest link in the chain—is called the limiting
amino acid. The essential amino acids most often acting in a
limiting capacity are methionine (and cysteine), lysine, and
tryptophan.

The table (p. 31) shows the percentage adequacy (or score)
of various proteins compared to reference protein. The best
quality proteins are those in which there are no limiting
amino acids. For instance meat, fish and eggs have a score of
100 and are therefore as adequate as the theoretical reference
protein. They contain all the essential amino acids in the
assumed right proportion and when eaten alone can meet the
body needs for protein repair, provided the diet also contains sufficient energy. Peanuts, however, have a limiting
amino acid—lysine—which is only present in approximately
65% of the right concentration. They have a score of 65 and,
in order to fulfil the day's needs for all the essential amino
acids, 100/65 or $1\frac{1}{2}$ times as much of peanut protein
would have to be eaten when it is the sole protein in the
diet.

Biological tests (mainly on animals—human tests are
rarely feasible) which measure directly the percentage of
protein retained in the body give slightly different values.
For instance, meat and fish proteins are not used with 100%
efficiency when fed as the sole protein in the diet. Egg has a
higher value than meat and fish. However, animals differ in
their needs for essential amino acids: both estimates can
therefore only be used as a guide for the relative value of one
protein to another.

A poor protein can be improved if it is eaten together (but
not at different meals) with some other proteins, when the
different proportions of essential amino acids may supplement each other. For instance, bread is limited by lysine—
only half the protein in wheat is used to make new body
protein when eaten alone. If, however, bread is eaten with
cheese, which is a good source of lysine, the combined pro-

teins are used with greater efficiency. Another supplementary combination is rice and pulses.

The protein in British diets, which contains a mixture of good, moderate (potato and wheat) and poor (leafy vegetables) quality proteins, is used (according to biological tests) with about 70% efficiency. It is limited by the amino acids methionine and cysteine. The minimum protein requirements are accordingly increased by a factor of 100/70 or 14%. (See Protein.)

QUALITIES OF VARIOUS PROTEINS

Protein	Limiting amino acid	% adequacy[a] compared with reference protein[b]	Biological tests[c]
Egg	None	100	94
Cod	None	100	83
Milk, cows	S aa[d]	95	82
Beef	None	100	70
Cheese	S aa	98	70
Soya beans	S aa	74	65
Rice, polished	Lysine	67	59
Wheat, whole	Lysine	53	48
Maize	Lysine	49	52
Peas	S aa	57	50
Potatoes	S aa	54	60
Peanuts	S aa	65	47
Gelatine	Tryptophan	0	2

a—or score, calculated from Amino Acid Contents of Foods and Biological Data on Proteins, FAO 1970

b—using new scoring pattern—FAO/WHO 1973, Energy and Protein Requirements, WHO Tech. Rep. Ser. 522

c—Net Protein Utilisation, taken from reference a. Values for Biological Value—which do not take into account undigested protein—are slightly higher

d—Sulphur amino acids—methionine and cysteine

In an effort to benefit some underdeveloped countries, whose protein is practically all derived from cereals, new high lysine strains are being bred. One possible disadvantage of these new strains is an imbalance of amino acids. In the USA, the quality of wheat protein is improved by the addition of lysine to enriched white bread. Similar proposals have not been accepted in Britain because lysine is not the limiting amino acid in British diets.

Synthetic diets, modified in their content of one or more

amino acids, are used in the treatment of several **inborn errors of metabolism.**

ANAEMIA

Insufficiency of the red pigment, haemoglobin, necessary to transport oxygen in the blood stream from the lungs to all parts of the body. Sufferers from anaemia are tired, breathless and pale. Severe anaemias are life-threatening. Often caused by a lack of **iron** in the diet, but see below.

Haemoglobin, composed of iron and **protein**, is contained in the red blood cells. Normal individual red blood cells circulate for about four months, when they are withdrawn and dismantled in the spleen. They are constantly replaced by new red blood cells synthesised in the bone marrow.

One cause of anaemia is lack of the **nutrients** required for normal renewal of blood cells. This may be brought about by a defective diet; by excessive blood losses (for instance because of heavy periods or peptic ulcers); inadequate absorption of nutrients from food (see malabsorption); and unusual needs, for example during pregnancy. Additionally there are many medical causes of anaemia, including inherited defects and kidney failure.

Nutrients particularly needed for red blood cell formation are iron, protein, vitamin B_{12}, folic acid, pyridoxine (vitamin B_6), copper and vitamin C. Depending on the nutrient lacking, too few cells, or cells containing insufficient haemoglobin may be formed.

In the UK, iron deficiency anaemia is the most common. Anaemia due to lack of folic acid is one complication of pregnancy, but diets inadequate in one of the other nutrients alone are rarely eaten—see individual entries.

ANOREXIA NERVOSA
See Starvation.

ANTIBIOTICS

Legally allowed as a preservative in fish and to enhance weight gain of pigs and chickens. The permitted antibiotics are also used in medical practice. Nisin, not used in medical practice, is a permitted preservative in all canned foods, cheese and clotted cream.

Fresh and frozen fish is permitted to contain a maximum of 5 parts per million of tetracycline (5 milligrams per kilogram of fish). One milligram of tetracycline would be supplied in a 200 gram (7 oz) portion of fish.

Antibiotics are usually added to the feed of pigs and chickens (though not laying hens): more edible food is produced from the same amount of feed as a result. Small amounts (less than 1 part per million) are left in the carcass: 1 kilogram of meat would have to be eaten to consume 1 milligram of antibiotic. When used medicinally, the daily dose is at least one gram (1000 milligrams). Antibiotics cannot be used to improve growth in sheep and cattle, which depend on bacteria in their rumen to digest **cellulose** in grass and feed.

It is possible, though unlikely, that people who have become allergic to antibiotics used to treat an infection, may be allergic to poultry, chicken or fish treated with antibiotic. Indiscriminate use of antibiotics to prevent infection in cows could result in significant quantities in milk. However, there is a more serious possibility: small pervasive amounts of antibiotic on the farm and in food could result in the multiplication of resistant strains. A subsequent infection would not then be so amenable to antibiotic treatment.

ANTIOXIDANTS—see also ADDITIVES

Added to fats to prevent rancidity. They prolong shelf life by conserving flavour but—in contrast to **preservatives**—do not retard putrefaction.

Rancid odours are due to compounds formed when fats

react with oxygen. Ultra-violet light, heat, and minute quantities of **copper, nickel** and **molybdenum** initiate rancidity. The impurities can be inactivated by sequestering agents (see Additives)—like **citric acid**—but heat (from cooking and in a warm storage area) and ultra-violet light (penetrating clear packaging materials) cannot always be avoided.

Several hundred chemicals have antioxidant properties, including **vitamins C** and **E**. The Antioxidant in Food Regulations 1974 permit the addition of those vitamins, or their salts, or synthetic equivalents, to any food. Vitamin C for instance is used in sausages and foil-packed vegetables.

Additionally, six other substances are permitted in specified foods up to specified amounts. All edible fats (except retail **butter**) and oils are allowed to contain BHT (butylated hydroxytoluene), BHA (butylated hydroxyanisole) or gallic acid esters. Apples and pears may contain ethoxyquin: it prevents browning, or 'scald' of the skin and flesh during storage. The other foods which may contain antioxidants are vitamin and flavouring (essential) oils and some **emulsifiers**.

Although the use of antioxidants is apparently restricted, they are carried over in any food in which fat, oil, or butter is an ingredient. Examples are crisps, frozen fried food, biscuits and cakes. Baby foods are exceptional: no ingredient containing antioxidant (other than vitamins E and C) is allowed to be used in their preparation.

Antioxidants are thought to be safe, but accumulate in the body fat (adipose tissue). Americans reportedly have accumulated more antioxidant than Britons—a probable reflection of their higher consumption of margarine and **convenience foods**.

ARGININE

An **amino acid**, needed for the growth of many young animals, but apparently a non-essential amino acid for children and adults. It occurs in all food **proteins** and in common with other amino acids, daily excesses are converted to **glucose** and used for **energy**.

ARROWROOT

A refined starch, produced from the roots of a West Indian plant. Sometimes called a **cereal** but it contains no germ, and is of poor nutritive value, providing **energy** but hardly any essential **nutrients**. Arrowroot is practically all digested, which probably accounts for its reputation as a household remedy for diarrhoea.

AVERAGE NUTRIENTS IN 15 GRAMS
(2 HEAPED TEASPOONS) ARROWROOT

Energy (kilojoules)	225
(kilocalories)	55
Carbohydrate (grams)	14
Protein (grams)	0·1
Fat	Trace
Calcium (milligrams)	1
Iron (milligrams)	0·3
Vitamins	0

ASCORBIC ACID

Chemical name for **vitamin C**. Large quantities of synthetic ascorbic acid are used in processed foods.

Ascorbic acid combines with oxygen and is used as an **antioxidant** in fats and meat, for example sausages, and frozen fish. It is not permitted in fresh and fresh minced meat. It also enhances the effect of **nitrite**, allowing less to be used in cured meats like ham. Significant amounts remain in these foods. In vegetables it prevents browning, and in fruits, fruit juices and carrots it stabilises the colour and preserves carotene (pro vitamin A). In **flour** ascorbic acid is used as an improver, especially in new **bread** making processes. In beer and wine it preserves colour and flavour, and in condensed milk it preserves flavour and vitamin A.

ASH

The **mineral** matter, left after foods are burnt in oxygen. All
other components—**carbohydrate, protein,** etc.—are removed
as gases.

May also refer to acidity or alkalinity of urine, which is
affected by diet. Diets high in **phosphorus** and **sulphur,** which
are excreted in the urine as sulphuric and phosphoric acids,
cause the urine to be acid. Diets high in **sodium** and **potassium** have an alkaline effect. Diets high in meat are acid ash:
diets high in fruits and vegetables are alkaline ash. Of little
relevance in normal health, since the kidney readily eliminates waste materials from the body.

ATHEROSCLEROSIS

The progressive accumulation of deposits of fatty substances **(lipid)** and fibrous tissue in the lining of the arteries,
which become thickened and constricted and eventually may
ulcerate. Blood tends to stick and form a clot (thrombus)
more easily in these areas, obstructing the flow of blood
through the artery. These blockages can occur in any large
or medium sized artery in the body, but they have more
serious consequences in arteries supplying the heart and
the brain.

Atherosclerosis with subsequent blockage of the blood
supply to parts of the brain is one cause of strokes. A high
blood pressure increases the risk of strokes—both from
atherosclerosis and other causes. **Obesity** and a high **salt**
intake are associated with high blood pressure.

There is much evidence that diet may influence atherosclerosis and resultant **heart disease**—the most common
cause of death in the UK—but there are other important
contributory factors, including high blood pressure, smoking, and **diabetes.**

BALANCED DIET

A diet which prevents deficiency diseases and which sustains a healthy and vigorous life. Generally taken to be a diet which contains all the essential **nutrients** in the optimum proportion for the individual. For practically all otherwise healthy people in Britain, a diet which supplies the **recommended intake** of each nutrient is usually assumed to be balanced.

Thus, a diet containing bread and margarine alone is not balanced: eventually deficiency diseases—notably **scurvy**—would develop. Although the addition of **potato** would improve it, the diet would still contain no vitamin B_{12}, and little riboflavin (vitamin B_2). The inclusion of **cheese** considerably improves the balance, but over the long term, even this diet carries with it the risk of iron or other trace element deficiency.

Most foods contain a variety of nutrients, but nearly all are deficient in one or more. At the same time, some food constituents enhance, and others hinder, the absorption of other nutrients. Additionally foods vary tremendously in their content of nutrients—particularly minerals and vitamins. Soil, climate, breed, degree of maturity, storage, processing, and cooking all cause variation in nutrient content. For this reason, analyses of foods should be treated with caution: they are average values.

In general, it is safe to assume that most people will achieve balance if a mixed diet—a wide variety of foods in moderation—is eaten. To simplify matters, foods can be divided into groups, according to the main nutrients they supply. A balanced (mixed) diet will be eaten if a number of portions is taken from each group every day. Within each group as great a variety of foods as possible should be eaten. Sugar and sugary foods—like soft drinks, and confectionery—and **alcohol** can be excluded without any risk of unbalancing the diet: like other foods they contribute **energy** but, unlike other foods, little else. See table overleaf and individual food entries.

For a normal adult man a balanced diet would consist of
two or more portions from the group of foods containing the
most protein (group 1) and three or more portions of potatoes, vegetables or fruit (group 2) each day. If no liver is
taken, green or yellow vegetables and fruits should be chosen

FOODS GROUPED ACCORDING TO THE NUTRIENTS THEY SUPPLY

Group	Foods	Best source of	Notes
1	Meat, fish, eggs, cheese, milk, yoghurt Pulses (peas, beans, lentils) Nuts	Protein Meat is best source of iron	Liver good source of vitamin A. Also contains vitamins B and E and other minerals
2	Vegetables Fruits Potatoes	Vitamin C	Vitamin A only in green and yellow fruits and vegetables. Also contain vitamins B and E, minerals and *fibre*
3	Milk, cheese, yoghurt	Calcium	Also all other vitamins and minerals but little iron and usually no vitamin C
4	Cereals	Energy, protein, B vitamins, fibre	Wholegrain cereals better than refined. Also contain minerals
5	Butter, margarine Other fats and oils	Energy Butter, margarine vitamin A and little D and E	No other vitamins and minerals. Oils are usually good source of vitamin E
6	Sugar, refined starches Alcohol	Energy	Little or no other nutrients

in preference to others. Fats (group 5) should be eaten in
moderation and cereals (group 4) should be eaten to satisfy
appetite and maintain a constant body weight. Some animal
protein food should be included.

Other groups of people have special needs:
Women of child-bearing age require more **iron** and should
include at least one portion of meat or fatty fish in the daily
diet.
In **pregnancy and breast feeding** more of all nutrients are

required to ensure adequate supplies for the growing child. The inclusion of extra **milk** or cheese (group 3) most easily meets these extra needs, together with plenty of meat and vegetables. Extra **fibre** should also be taken to avoid constipation.

Children and adolescents require more nutrients in proportion to their weight than adults to allow for growth. Extra milk most easily meets these needs. If there is little exposure of the skin to sunlight, extra **vitamin D** (from cod-liver oil, margarine and fatty fish) is required.

The elderly require less energy but the same quantity of all other nutrients. Foods high in energy, but poor sources of essential nutrients—like fried foods, sugar, cakes, pastries and biscuits—are best avoided. Tinned meat, fish, vegetables and eggs are easily prepared and chewed sources of essential minerals, protein and vitamins. A $\frac{1}{2}$ pint of milk (or dried or condensed) should be taken each day, together with extra **vitamin D** if there is little exposure of the skin to sunlight.

In the UK, the average intake of each of the main nutrients is greater than the average recommended intake—see table—except for vitamin D, which is also synthesised in the skin when exposed to sunlight. Although in theory the average diet is balanced, averages mask individual intakes and needs and deficiency diseases do occur. Additionally, 'modern' nutritional diseases—obesity and attendant complications, dental caries, and probably late onset diabetes and heart disease—are serious public health problems—see below.

Deficiency diseases are particularly likely to affect the elderly and exacerbate mental confusion and listlessness. They most commonly suffer from **osteomalacia** due to insufficient vitamin D in the diet and from synthesis in the skin. Through infirmity, they are often unable to cook and eat sufficient protein and vegetable foods, which supply iron, vitamins B and C and **potassium**. Anaemia is therefore also common and skin complaints and scurvy occasionally occur.

The other groups most at risk—infants, children, adolescents and pregnant and breast-feeding mothers—are largely protected by the welfare state, which supplies nutritional supplements in pregnancy and free milk and school meals if the diet is likely to be prejudiced by lack of money. Rickets, however, occurs more frequently in Asian children—see vitamin D.

In adults, nutritional deficiencies are rare, except for iron deficiency anaemia in women (see Iron). Occasional deficiencies result from fad diets, but fortunately the body can cope with unbalanced diets for short periods. Most people cannot tolerate one food—to the exclusion of others—for long. Other causes of deficiency diseases in adults are drug addiction (for instance beriberi in alcoholics), medical disease (osteomalacia and anaemia in **malabsorption**) and restriction of the diet for medical purposes (scurvy in milk-based ulcer regimes and vitamin A deficiency from a diet used for self-treatment of epilepsy).

Modern nutritional diseases are almost endemic in industrialised societies. Technology reduces activity at home and work, and although less energy is required in the diet, it also ensures lavish supplies of food. **Obesity** and its attendant complications become almost inevitable.

Other changes brought about by industrialisation affect the type of food eaten. Cheap filling starches (in the UK from bread and potatoes) are replaced by sugar and fats. They are concentrated sources of energy and lacking in other nutrients compared with bread and potatoes. The consumption of **convenience foods** also rises: these supply as much energy but are often lacking in essential nutrients compared to traditional alternatives prepared in the home. Both these changes contribute to obesity, or—since the need for all other nutrients remains the same—may result in a diet containing the correct amount of energy but lacking in essential nutrients.

To some extent these changes are counteracted in industrialised societies because more expensive animal foods (meat, eggs and milk) can be bought. These replace the

NATIONAL AVERAGE INTAKES OF NUTRIENTS COMPARED WITH
RECOMMENDED INTAKES FOR 1973 *

Nutrient	Average intake per person per day	% adequacy compared to average recommended intake
Energy	2400 kilocalories (10,000 kilojoules)	104
Protein	71·4 grams	124
Calcium	1020 milligrams	193
Iron	12·7 milligrams	118
B vitamins:		
Thiamin	1·22 milligrams	134
Riboflavin	1·79 milligrams	132
Nicotinic acid	29 milligram equivalents	193
Vitamin C	53 milligrams	189
Vitamin A·	1270 microgram equivalents	190
Vitamin D	2·89 micrograms	89

* Taken from Household Food Consumption and Expenditure 1973
HMSO

vitamins and minerals lost from starchy foods and deficiency
diseases are rare. However, although the diet supplies the
recommended intakes for all the main nutrients, other
modern diseases—**heart disease**, dental caries, gall stones and
diverticulitis (see Fibre)—may be due to a diet that has
become 'over-balanced': it contains too much fat and sugar
and possibly too little of the lesser known nutrients.

BARLEY

Similar nutrient value to other **cereals** with most of the
minerals and vitamins in the outer layers. Hulled (Scotch)
barley retains all the **nutrients** in the grain, but pearl barley
has the outer bran and germ removed, when 70 % of **thiamin**
(vitamin B_1) is lost.

Barley meal (from hulled barley) and barley flour (from
pearl barley) can be made into bread, but wheat is preferred
because of its higher content of **gluten**. Barley water is made
from pearl barley boiled in water—it contains 2 % **starch** but
little else (see soft drinks).

Two-thirds of the UK grain crop is barley but little is now used for human consumption. Nearly 90 % is fed to animals. The rest is used to make malt (see maltose) used in **beer**, vinegar, whisky, malted drinks, malt flour and for malt extract with cod-liver oil.

AVERAGE NUTRIENTS IN 25 GRAMS (1 OZ) UNCOOKED BARLEY

Nutrient	Hulled	Pearl
Energy (kilojoules)	355	370
(kilocalories)	85	90
Protein (grams)	3	2¼
Fat (grams)	½	⅓
Carbohydrate (grams)	17	19
Calcium (milligrams)	10	5
Iron (milligrams)	1	0·2
B vitamins:		
thiamin (milligrams)	0·13	0·04
riboflavin (milligrams)	0·05	0·02
nicotinic acid (milligram equivalents)	0·75	0·55
Vitamin B_{12}	0	0
Others	—	Less than hulled
Vitamin E	—	Less than hulled
Vitamins A and C	0	0
Other minerals and trace elements	Small amounts	Less than hulled

— – no information available

BEER—see also Alcohol

Beer contains alcohol and sugar, both of which supply energy. Most beers contain from 75 to 120 kilo**calories** per ½ pint; from 6 to 13 grams **alcohol** per ½ pint; and from 4 to 13 grams carbohydrate per ½ pint.

Beer is a good source of some B vitamins (see table) but poor in thiamin (vitamin B_1) which is retained by brewer's **yeast**. It contains no vitamins A or C, virtually no protein, and little minerals, except **potassium**.

Beer is among the least alcoholic of the alcohol containing beverages—most have 2–4 % alcohol by weight—but it is drunk in the greatest quantities in Britain. The average adult consumption is about 4¾ pints per week in the UK. Beer is a general term for ale, stout and lager, but it is also used synonymously today with ale, which was originally brewed in the medieval home from any cereal. Beer eventually became the more popular drink after the introduction of hops from Flanders in the fifteenth century.

The differences in ale, stout and lager are due to modifications of the basic process: fermentation of sugars with yeast.

AVERAGE NUTRIENTS IN 300 MLS (½ PINT) BEER

Nutrients	Bottled		Draught		
	Pale ale	Stout	Mild	Bitter	Strong
Alcohol (grams)	10	8	8	9	20
Energy (kilojoules)	405	465	315	390	920
(kilocalories)	95	110	75	95	220
Protein (grams)	1	1	½	¾	2
Carbohydrate (grams)	6	12	5	7	18
Potassium (milligrams	145	135	100	120	330
Other minerals	Small amounts				
B vitamins:					
thiamin (milligrams)	.... average 0·012 ...				0·03
riboflavin (milligrams)	.... average 0·15 ...				0·3
nicotinic acid (milligram equivalents	.... average 2·1				4·5
pyridoxine (milligrams)	.... average 0·18				
pantothenic acid (milligrams)	.... average 0·3				
biotin (micrograms)	.... average 1·5				
others	—	—	—	—	—
Vitamins A and C	0	0	0	0	0
Vitamin E	—	—	—	—	—

— - no figures available

Barley is the common cereal and it is malted by allowing it to sprout when **enzymes** partially convert the starch to **maltose** and **dextrins**. After drying, the sugars are dissolved out of the grain in water—to form wort—boiled with hops and inoculated with a pure culture of brewer's yeast. Resins in the hops flavour and also preserve the beer by inhibiting the growth of acetic acid forming bacteria.

Bitter beers contain the most hops. For stout brewing, the malt is dried at a higher temperature, when the grain is partially charred. Stout has a higher carbohydrate content than most beers, extra stout more alcohol. Lager contains less alcohol than beers or stout but it is absorbed more quickly into the blood stream. A different strain of yeast—which ferments at the bottom of the vat—increases the effervescence of lager. Strong ale is the most alcoholic—6% by weight—and contains more energy (about 220 kilocalories per $\frac{1}{2}$ pint), B vitamins, minerals and carbohydrate than all other beers.

BERIBERI

A nutritional disease, caused by a diet containing insufficient vitamin B_1 (**thiamin**). It is rare in Britain, except amongst alcoholics, but used to be endemic in communities where the diet was lacking in variety and based on highly milled unfortified **cereals**. It has not been completely eradicated from some **rice**-eating countries because it is difficult in rural areas to enforce either parboiling of the rice before milling or subsequent replacement with synthetic vitamin B_1. The vitamin is replaced in all British white wheat **flours** and some refined breakfast cereals.

Beriberi is primarily a disease of the brain and nerves, which are more dependent on thiamin than other tissues. An initial general lassitude and loss of appetite is followed by impairment of skin senses and memory. Walking becomes difficult and work potential falls as the nerves degenerate and associated muscles atrophy. An infection or extra stress such as hard labour may precipitate death.

The thiamin content of breast milk reflects the dietary content, and infants may fall ill from beriberi even though the mother is seemingly in moderate health. The child may appear well fed, due to the accumulation of water in the tissues, but unless thiamin is supplied, is liable to die suddenly from heart failure.

BEVERAGES

Essentially flavoured **water**, of which most people require at least $1\frac{1}{2}$ pints to drink a day, and more in a hot climate.

The flavours in beverages are derived from essential **oils** in berries, fruits, herbs and grains. Some are valued chiefly for their **caffeine** and **alcohol** content. **Soft drinks** and alcoholic drinks contain **energy** and all except spirits may contain small quantities of minerals and vitamins. See individual entries: beer, cocoa, coffee, mineral waters, soft drinks, spirits and tea.

BIOFLAVONOIDS—including citrin and rutin

Derivatives of the flavones—widely distributed plant pigments (see colour). Rutin is found in cereals; citrin in citrus fruits, especially the pith.

One of the symptoms of **scurvy** is haemorrhage due to leakage of blood from the walls of small blood vessels (capillaries)—noticed as bruises under the skin. In guinea pigs, natural sources of **vitamin C**, which contain bioflavonoids, sometimes heal the blood vessels more effectively than vitamin C alone.

The active substance was isolated from peppers in 1936 and named vitamin P, but since the bioflavonoids have not been proven as essential nutrients for animals, man or microorganisms, they are not strictly a vitamin. They strengthen capillaries weakened by lack of vitamin C probably by suppressing the formation of substances, including histamine (see histidine) normally released when blood vessels are damaged.

Bioflavonoids were once claimed as a cure for the common cold, but later experiments have shown that they have no effect. They are unlikely to be lacking in diets which contain vitamin C.

BIOTIN

Part of the **vitamin B complex**. It is an important coEnzyme, necessary for many vital body processes, but, except under very unusual circumstances (see below), adult deficiency is unknown. Most daily diets contain 150–300 micrograms, but bacteria normally present in the intestines synthesise at least 2–5 times this amount, some of which is absorbed into the blood stream. The vitamin is widespread in food (liver, kidney, egg yolk and yeast are rich sources) and it is unaffected by cooking and processing. Breast milk contains very little: infants suffering from diarrhoea occasionally become deficient in biotin.

Egg white contains a **protein**, avidin, which combines with biotin, making it unavailable for absorption into the blood stream. However, in cooked eggs, heat inactivates avidin and the availability of biotin is unimpaired. One or two raw eggs a day are harmless, but large quantities of raw egg white inactivate both the vitamin in food and that synthesised in the intestine. A diet taken for 3–4 weeks by volunteers, containing the equivalent of 80 raw egg whites a day, resulted in a variety of symptoms—including nervous disturbances and skin lesions (seborrhoeic dermatitis)—all of which were cured with injections of 100–300 micrograms of biotin. Some freak diets, for instance one dozen raw eggs each day, have also occasionally resulted in biotin deficiency.

Excess intakes of biotin are harmless and eliminated from the body in urine. In the treatment of skin disorders large doses are of dubious value, except when deficiency has been induced with egg white.

Animal experiments suggested that induced deficiency of biotin might promote regression of some cancers, but raw egg white diets were found to be of no value when used to treat human cancer.

BLAND DIET
See Dyspepsia.

BRAN
Probably the best source of dietary **fibre**. It is the outer husk of **wheat**, weighing about 13 % of the whole grain. About 20

grams (6 teaspoons) of bran, taken with milk, fruit, soup or yoghurt, each day are recommended for the treatment of constipation. An equivalent amount of fibre is contained in about 120 grams (5 to 6 small slices) of wholemeal **bread**, or 40 grams (2 to 3 normal helpings) of 'All-bran'.

BREAD—see also Flour, Wheat and Cereals

An important source of **energy, protein** and B **vitamins**. It contains small amounts of **minerals**—including trace elements—but no vitamins A, D, B_{12}, or C. In Britain white bread is a significant source of **calcium**.

There are three main types of bread made: white, brown and wholemeal. They are mainly **starch** and protein but differ in their content of other **nutrients** and, see below, additives.

Wholemeal bread (sometimes called wholewheat) is made from wholemeal flour containing all the constituents of the wheat grain—including the germ (rich source of B vitamins, vitamin E and minerals) and the **bran** (rich source of fibre). It contains an average of 2% **fibre**.

White bread is made from white flour from which all the bran and germ—about 30% by weight—have been removed. White flour contains about 0·2% of fibre. White bread is therefore a poorer source of fibre and contains less B vitamins, trace elements and vitamin E than wholemeal. To overcome some of the losses in milling, 3 nutrients are partly replaced in white flour—see below.

Brown bread is made from brown flour with a fibre content intermediate between white and wholemeal flours. The legal minimum fibre content of the flour is 0·6%—that of a flour of 85% extraction (see flour). It can be produced by milling off 15% of the wheat grain (containing most of the bran and some of the germ): in this case the nutritive value is intermediate between white and wholemeal. Alternatively brown flour can be a blend of white flour with sufficient bran to comply with the legal requirement; apart from fibre, the nutritive value is then approximately that of white flour. Wheatmeal breads are used to describe a bread with a higher fibre content than the legal minimum, but the Foods Standards Committee has recommended that wheatmeal should not be a permitted name in the future. It is too easily confused with wholemeal and does not legally have to contain any germ. See also wheat germ and granary breads below.

Wholemeal flour must not contain any added nutrients, but to overcome some of the losses in milling, the nutrients **iron** and two B vitamins (**thiamin**, vitamin B_1, and **nicotinic acid**) are replaced by law in white flour to the level of 80% extraction (see flour). Blended brown flours (white flour with bran) which do not contain sufficient germ, must also contain these added nutrients to comply with the legal requirements (though in practice, millers usually add the nutrients to all brown flours and most brown bread contains more thiamin than wholemeal). However, it is now realised that most of the natural nicotinic acid in cereals is in a bound form, not available to man, and consequently brown and white flour contain more nicotinic acid equivalents than wholemeal. To overcome this anomaly the Food Standards Committee (FSC) has recommended that the addition of nicotinic acid should be discontinued. **Calcium** carbonate (chalk) must also be added to all flours except wholemeal— see Phytic acid.

White bread has always been preferred to wholemeal, originally because it could be eaten without expensive butter or other foods. Additionally, bakers prefer white flour for baking: bran and substances in the germ cause a heavier and darker loaf. White (wheaten) bread was a luxury food, available only to the richer classes, before the introduction of roller mills (see also flour). Today, about 80% of all bread eaten in the home is white and less than 2% wholemeal.

White bread is a nutritious food. The protein and energy are more completely digested than in wholemeal and white bread contains virtually no **phytic acid** which interferes with the absorption of some trace elements and calcium. The

Average nutrients in 25 grams (1 thin slice) bread

Nutrient	Whole meal	Brown	Added wheat germ[a]	White
Energy (kilojoules)	240	255	250	255
(kilocalories)	55	60	60	60
Protein (grams)	2	2	2½	2
Fat (grams)	½	½	½	½
Carbohydrate (grams)	12	12	12	13
Calcium (milligrams)	5	25[b]	25[b]	25[b]
Iron (milligrams)	0·72	0·6	0·67	0·45[b]
Sodium (milligrams)	115[c]	140[c]	130[c]	130[c]
Potassium (milligrams)	65	60	60	25
Magnesium (milligrams)	20	15	15	5
Copper (milligrams)	0·12	0·09	0·09	0·04
Other minerals and trace elements	Wholemeal more than other breads			
Vitamins A and C	0	0	0	0
B vitamins:				
thiamin (milligrams)	0·05	0·05	0·07	0·05[b]
riboflavin (milligrams)	0·02	0·02	—	0·01
nicotinic acid (milligram equivalents)	0·5	0·65	—	0·7[b]
pyridoxine (milligrams)	0·08	—	—	0·01
pantothenic acid (milligrams)	0·18	—	—	0·13
folic acid (micrograms)	—	—	10	1·5
biotin (micrograms)	0·5	—	—	0·13
vitamin B_{12}	0	0	0	0
Vitamin E (a tocopherol milligrams)	0·03	—	—	Trace

a – for example Hovis
b – added by law
c – added during bread making. Salt-free bread contains little sodium
– – no figures available

added chalk makes white bread a significant source of calcium in the diet. However, complete removal of the bran and germ from the flour causes severe losses of nutrients. Apart from those partially replaced other reported losses include approximately 70% of pyridoxine (vitamin B_6), 80% of riboflavin (vitamin B_2), 50% of pantothenic acid, 85% of vitamin E, 65 to 85% of the trace elements copper, manganese, zinc and of magnesium and 90% of fibre. The small amount of vitamin E remaining in white flour is destroyed by bleaching agents (see Flour).

The relevance of the loss of these nutrients to health when bread is eaten as part of a normal **balanced diet** is uncertain. Doubts about the adequacy of fortified white bread, compared with wholemeal or brown, were apparently contradicted by the results of a trial in 1947 which demonstrated that children grew equally well when fed wholemeal, brown, unfortified and fortified white breads, provided the diets were otherwise satisfactory as regards iron, calcium and vitamins A, D, and C. However, some of the discarded nutrients may be protective factors against **heart disease**, which occurs in adult life. See also fibre. It should be noted that wholemeal bread requires more butter, jam and other spreads to make it palatable. Excess consumption of these foods may be as harmful as the loss of nutrients in white bread.

In view of the probable importance of cereal fibre to health in modern societies, it is unfortunate that there is much confusion about the difference between wholemeal and brown breads—even amongst bread shop assistants. Very little wholemeal bread is available in bakery shops and supermarkets. The abolition of the term 'wheatmeal' should partially clarify the situation. Additional information could be made available through the Flour Advisory Bureau (financed by the milling and bakery industries). At present most of its advertising resources seem to be devoted to selling white bread.

The quantity of bread eaten in the home has been declining steadily over the past 20 years. In 1956 the average household consumption of all bread was approximately 50 oz per person per week (or about $5\frac{1}{2}$ large slices per day). This had fallen to approximately 33 oz per person per week (or about 4 large slices per day) in 1973. The main reasons for its decline in popularity are a better standard of living and a general tendency towards a more sedentary life. These two factors have contributed to the replacement of starchy filling foods (like bread) with more expensive meals containing more animal protein and fat. Bread has also an unfortunate reputation as a fattening food and it is automatically excluded when the diet is changed in an attempt to lose weight.

Bread is still an important staple food in Britain, despite its decline in popularity. The amount eaten outside the home—in snack bars at lunchtime for example—has probably increased over the past 20 years. Even in the average household diet, however, it still supplies approximately a quarter of the thiamin (vitamin B_1), a sixth of the protein and a seventh of the energy (as starch). It is of more importance in low-income large families who are more dependent on bread as a cheap source of food.

The belief that bread is higher in **energy** than most other foods is unfounded. Although about half its weight is starch, bread contains little fat (white contains virtually none) and over a third (35–38%) of its weight is water. A large slice of bread (about 35 grams) contains the same **energy** as an egg (90 kilo**calories**). Some bread can be included in slimming diets (see Obesity) but care should be taken with butter and margarine used for spreading—see 'slimming' breads below.

Staling of bread is due to chemical changes similar to those which take place when starch sauces and puddings set. Part of the starch molecules become linked closer together, squeezing water out of the middle of the loaf. The crumb becomes dry and the crust soft and tough. In the early stages, staling can be reversed by heating the bread in an oven.

Toasting of bread drives off some of the moisture (when used in a slimming diet bread should therefore be weighed before toasting) and browns the surface by causing caramel and **dextrins** to be formed. Toast bread contains more sugar than normal bread so that browning is achieved in a shorter time. Toasting destroys up to 30% of thiamin (vitamin B_1) in bread. More is lost from thin than thick toasted slices.

The flavour of bread is mainly due to yeast, salt and the crust. In the heat of the oven, caramel, and a reddish brown complex are formed, colouring the crust. The protein in the crust has a slightly lower nutritive value than the crumb: amino acids in the colour complex (see Colour) are inactivated. Crusty loaves are cooked at a lower temperature and for a longer time than slicing ones. Although bread is not actually steamed or baked in steam, the high temperature needed for slicing loaves is achieved in ovens heated with steam-containing pipes (steam baked).

Food additives used in bread

Bread used to be bought more frequently and from small bakeries. It was crusty and eaten fresh. Sliced wrapped bread—a **convenience food**—must be cooled before slicing: its relative staleness and lack of crust (to facilitate slicing) contribute to its reputed lack of flavour and rubbery texture. Additionally the decline in popularity of bread, together with reduction in the time spent shopping and less frequent purchases, have contributed to the replacement of the small baker with large bread making concerns. About 70% of all bread is now made in large baking factories and over half of all bread is sold in supermarkets. Bread sold in this way must have a longer shelf life—to allow time for delivery—and, like other convenience foods, be a standardised product. **Additives** are used to achieve this aim.

The eight groups of additives permitted in bread are shown in the table opposite. Wholemeal bread may contain all of them but it is not permitted to contain, in addition, the bleaches and improvers used in brown and white flour (see Flour). Caramel (brown colour) is permitted in brown and wholemeal flour but not white. The FSC has recommended that two improvers, at present permitted only in white and brown breads, should be allowed in wholemeal bread in future. These improvers are **ascorbic acid** (vitamin C) and L cysteine hydrochloride (a salt of the amino acid **cysteine** naturally present in proteins including those in bread). These improvers are necessary for one of the new methods of bread making—see below.

Traditionally bread is raised (or proved) during a 3 to 4 hour period before baking. **Enzymes** in the flour and yeast cause the gas carbon dioxide to be formed which is trapped in the dough. During proving and kneading the protein

PERMITTED ADDITIVES IN BREAD

Type	Numbers permitted	Example(s)	Reason for use	Notes
Yeast stimulating preparations	5*	Ammonium chloride, Calcium sulphate	Mineral salts which promote vigorous growth of yeast	* Numbers allowed are not yet specified by law
Enzymes		Malt flour, Soya flour, Amylases	To overcome natural enzyme variation in different flours	* Added as malt or soya and more recently isolated from bacteria and fungi
Emulsifiers	3	Lecithin	See emulsifiers	
Preservatives	1	Sorbic acid	Prevents growth of mould	FSC have recommended that carbon dioxide be permitted
Colour	1	Caramel	Brown colour	Not permitted in white bread
Excipients	3	Starch	As dilutents for other additives	
Rope inhibitors	7	Vinegar, Lactic Sodium acid diacetate	Acids which prevent the growth of bacteria causing rope	'Rope' causes yellow-brown spots in the middle of loaves. In severe cases the loaf turns brown and can be pulled into threads
Antioxidants	5		Carried over from fats used—see Antioxidants	

Note. Bleaches and improvers (see flour) are permitted in all breads except wholemeal. Nutrients (see text and flour) must be added to all flour except wholemeal and some brown flours. Calcium (as chalk) must be added to all except wholemeal.

Milk, other cereals (as flours or wholegrains), seeds, gluten, and wheat germ may be added to white and brown breads but not wholemeal. Sugar is permitted in all bread.

gluten is stretched and made more elastic so that more gas can be held in the dough and a lighter loaf produced.

The Chorleywood Bread Process (CBP) is a new bread process that eliminates the need for proving. It allows bakers to use more British (weak) flours in bread—previously only strong imported flours, containing more gluten, were used to make the standard British loaf—and produces more loaves from the same quantity of flour. In traditional methods up to 7% of the weight of flour is lost during proving: alcohol is also formed by the yeast and is driven off from the dough during baking. The gluten fibres are stretched by a few minutes of intense mixing with high powered machinery. Improvers (see Flour) are necessary. About three-quarters of all bread eaten is now made by this process but it is largely confined to large bread making plants that have the capital available for the high powered equipment. CBP bread has a slightly lower protein content than normal bread—because of its higher proportion of British wheat—but otherwise it has the same nutrient composition.

Activated Dough Development (ADD) is another new bread making process which also eliminates the need for proving but does not require high powered machinery. The required changes in the structure of the gluten are produced by a combination of the improving agents ascorbic acid (usually together with potassium bromate) and L cysteine hydrochloride. At present the method is not widely used because L cysteine hydrochloride has only recently been permitted in bread, but it is expected to be used in small bakeries and for the production of speciality breads in large plants.

The flavour of breads made by these new methods is said to be indistinguishable from existing products but they have one possible disadvantage. Wholemeal bread made with new methods is likely to have retained nearly all phytic acid in the flour, which could interfere with the absorption of trace elements, particularly zinc. In traditional methods, the enzyme phytase, in yeast, had sufficient time to destroy about half the phytic acid during the long proving time. Most phytic acid is removed when wheat is milled to white flour and the difference in white bread is unimportant.

Special breads

Wheat germ bread must contain at least 10 % of added germ. When made from brown flour it contains more vitamin B_1 than wholemeal, but less fibre. Most types are made from brown flour, but some are made from white. It is subject to the same legal requirements as brown bread. The FSC has recommended that the description 'with added wheat germ' should replace 'wheat germ bread'.

Granary bread is a brown bread with added malted flour (see Maltose) and kernels of wheat or other cereals. It is not equivalent to wholemeal bread.

Soda bread is leavened with **raising agents** in place of yeast. It contains less vitamin B_1 than yeast leavened breads because alkalies destroy the vitamin. It can be made from brown, white or wholemeal flour and is subject to the appropriate legal requirements.

High protein breads or starch 'reduced'—contain added gluten or soya flour. This lowers the proportion of starch in the dough so that lighter (containing more air) breads can be made. They have virtually the same energy as ordinary bread. Ordinary bread contains about 13 % of protein (by dry weight); gluten breads (sold as 'light breads'—see Slimming breads below) contain about 16 %; high protein breads about 22 %; and starch reduced breads more than 32 %. The FSC has recommended that the name 'starch reduced' should not be permitted because it implies that such bread is less 'fattening' than normal bread. 'Extra high protein' should replace starch reduced as a description.

'Slimming' breads are advertised as starch reduced or lighter than normal bread—for example Procea, Nimble, Slimcea. They contain added gluten (and therefore proportionately less starch). Like other proteins, gluten has about the same energy value as starch (4 kilocalories per gram). Both these types of bread contain more air and are sliced thinly, so that *slice for slice* they do contain less Calories than normal slices of ordinary bread. However, when compared on an ounce for ounce basis, they contain at least as many Calories as normal bread. The FSC has recommended that 'light' breads should be subject to the same controls as other 'slimming' and starch reduced products—that is their Calorie content per ounce (or 100 g) should be stated and there should be a full declaration of ingredients on the wrapper. Ordinary bread is exempt from declaring ingredients.

Rye bread contains 20 to 60% of rye flour, together with citric acid (to give a sour taste), and sometimes caramel.

Part-baked bread. Bread which has the same composition as ordinary bread, baked at a lower temperature to 'set' the dough. Reheating in the oven has the same effect as it does in reversing staling in ordinary bread.

Bread for medicinal purposes. Salt-free bread is available from some bakeries. It has an inferior taste and stales more quickly than salted bread. Gluten-free bread is made from flour from which the gluten has been removed. It is improved by the addition of milk, or other gluten-free protein, to the flour. Both flour and ready-made bread are available on prescription for those suffering from **coeliac disease**, and many small bakers are willing to make special batches for customers. Low protein bread, needed for some types of kidney disease and phenylketonuria (see Inborn Errors of Metabolism), is made from gluten-free flour, but with no added protein. Both protein-free flour and bread are available on prescription for the treatment of these diseases.

Aerated bread. A method of raising bread by pumping the gas carbon dioxide into the dough. It is lacking in flavour compared with normal bread because no yeast is used. The FSC has recommended that gluconodeltalactone (a derivative of glucose) which liberates carbon dioxide slowly, leaving no after-taste (unlike soda bread) be permitted for use in aerated bread.

BREAKFAST

Children benefit from breakfast. They require more **nutrients** in proportion to their weight than adults, to allow for growth, and are more likely to fulfil their needs if they have 3—rather than 2—meals a day. Ideally, breakfast should supply about a quarter of the day's nutrients. For example, 2 slices of bread and butter, one egg, marmalade and a cup of milk supply about a quarter (550 kilo**calories**) of a 12-year-old girl's **energy** requirements for the day.

For adults, there is no evidence that breakfast is of benefit in promoting mental efficiency in the morning. One pilot study suggested that change in breakfast habits was more deleterious to mental performance than habitual lack of breakfast. In contrast, physical performance during the morning is probably improved by food before work—accident rates in steel workers have been lowered by a glucose drink (given for technical reasons) taken in the morning.

Exercise taken after a meal may help to avoid overweight. For this reason it is probably beneficial to eat a moderate breakfast and lunch, but avoid a heavy evening meal, when most people take little exercise. Meals containing 550 kilocalories contain sufficient energy for about 5 hours' office work (see Energy).

BUTTER

Concentrated milk **fat**. It is a rich source of **energy** and **vitamin A**. It also contains vitamins D and E. Summer-produced butter contains more vitamins than winter-produced. In contrast to other dairy foods, butter contains virtually no calcium, protein or B vitamins.

As in most other countries, the average consumption of butter has fallen over recent years in Britain—to just under $5\frac{1}{2}$ oz per person per week in 1973—and has been partly replaced by **margarine**. Butter contains **cholesterol**, which is absent from most margarine. Although avoided in cholesterol-lowering diets, butter in moderation is harmless for otherwise healthy people.

The natural **colour** of butter is due to **carotene** (from grass), but annatto or turmeric may be added to pale butter, especially in winter. The flavour of butter is mainly due to diacetyl and short chain **fatty acids**. Sweet butter, made from fresh cream, contains less diacetyl than ripened butter—made from cream soured with cultures of **lactic acid** producing bacteria. **Antioxidants** may not legally be added to retail butter, but butter sold for manufacturing or catering pur-

58 CAFFEINE

poses is permitted to contain them. Butter in—for example—
biscuits and cakes and butter served in individual packs in
restaurants will contain antioxidant. Up to 2% salt may be
added as a flavour enhancer and preservative: salted butter
may have to be avoided in **sodium** restricted diets.

AVERAGE NUTRIENTS IN 10 GRAMS (⅓ OZ BUTTER)

Nutrient	In 10 grams butter	Notes
Energy (kilojoules)	330	—
(kilocalories)	80	—
Fat (grams)	8½	—
Cholesterol (milligrams)	25	—
Protein, carbohydrate	Trace	—
Sodium (milligrams)	60	Salt-free butter only traces
Other minerals	Traces	Butter is poor in calcium
Vitamin A (equivalents)	105	—
Vitamin D (micrograms)	0·12	—
Vitamin E (α tocopherol, milligrams)	0·1	—
B vitamins	0	—

CAFFEINE

Contained in **tea, coffee** and **cocoa**. In mild doses, it is a
stimulant, decreasing reaction time and increasing mental
alertness, heart rate and the flow of urine.

One cup of tea contains 60–90 milligrams; one cup of
coffee 100–150 milligrams; and most cola drinks 40–70
milligrams per 11½ oz can (see Soft drinks).

Caffeine is harmless in small doses, but in excess it causes
sleeplessness, palpitation, tremor and depression. Over 600
milligrams—taken over a short period—is the toxic dose.

CALCIUM

Calcium is an important **mineral** in the diet. The bones—the
supportive framework of the body—are hardened with calcium absorbed from food during growth. It is also necessary

for tooth formation and for the normal activity of nerves and muscles.

Milk, most cheeses and yoghurt are the richest sources of calcium. Three-quarters of a pint of milk supplies the recommended intake for a child between the ages of 2 and 9. White and brown (but not wholemeal) bread, biscuits and other foods made from white **flour** are also good sources. There is virtually no calcium in butter, double cream, cream cheese and artificial whiteners. Some dark green vegetables, like watercress, are good sources, but spinach, and beet greens contain **oxalic acid** which renders most of the calcium unabsorbable. Wholegrain cereals, nuts and pulses contain **phytic acid** which also interferes with calcium absorption. Meat, fruit and fish (except fish with very small bones which are usually eaten) are poor sources. Hard water may add significant quantities of calcium to the diet, but soft water contains little. There is evidence that calcium in hard water may be beneficial, but it is not known if calcium itself is the protective factor—see Water.

The daily **recommended intake** for adults is 500 milligrams of calcium. The table below shows average portions of foods containing this amount. 1200 milligrams of calcium is

AVERAGE PORTIONS OF FOODS SUPPLYING THE ADULT RECOMMENDED
INTAKE OF CALCIUM

	Food	Approximate portion containing 500 milligrams
Rich sources	Most cheese, dried skim milk, whitebait, smelts, sprats	Less than 100 grams (3½ oz)
Good sources	Stilton, sardines, shrimps, fruit gums	100 to 150 grams (5 oz)
	Milk chocolate, watercress, parsley, pilchards, soya flour	150 to 250 grams (9 oz)
	Milk, ice cream, camembert, shellfish, herring, dried figs	250 to 500 grams (1 lb)
Moderate sources	White bread, cakes, biscuits, eggs, most dark green vegetables	500 grams to 1 kilogram (2¼ lb)
Poor sources	Cream cheese, butter, cream, fats, meat, white fish, most fruit, most vegetables, wholemeal bread	Over 1 kilogram

recommended during pregnancy and during breast feeding to meet the increased needs of the growing child and children require more in proportion to their weight than adults to allow for growth—see Appendix. The average British diet contains about 1000 milligrams of calcium, excluding water, of which over half (58 %) is supplied by milk and cheese, and a fifth (22 %) is supplied by cereals.

During **digestion** under the influence of **vitamin D**, calcium is absorbed out of food into the blood stream and transported to the bones. The protein (gristle) element is hardened with calcium and **phosphorus** during growth, so that by adulthood most people have accumulated about 1200 grams (approximately $2\frac{1}{2}$ lb) of calcium. A good intake of calcium is particularly important for children: not only may dietary insufficiency stunt growth, but a skeleton containing plenty of calcium may be protective against **osteoporosis** developing in middle age. Once growth has ceased, a diet high in calcium is probably ineffective in protecting against osteoporosis.

The majority of calcium is in bone, although a small but vital amount—about 10 grams—is held in the blood stream and cells of the body. A portion of this calcium regulates the activity of nerves and muscles, stimulates the secretion of some hormones, and assists in the clotting of blood. Together with vitamin D, two hormones—parathormone and calcitonin—delicately maintain the level of calcium in the blood stream.

Although no more bone is formed in adulthood, calcium is constantly withdrawn from the skeleton and, in health, replaced at an equal rate. A small but uncertain quantity is required in the diet for daily replacement. Usually the diet contains far more than daily needs and less than a third of that in food is absorbed into the blood stream. An equivalent amount is eliminated from the body by the kidneys in urine. If the diet is rather low in calcium (for example if no milk is taken) there is an adaptive response—provided there is sufficient available vitamin D—and proportionately more calcium is eventually absorbed from food. The adaptive response is less efficient with increasing age.

When the adaptive response, together with a reduction of the calcium eliminated in the urine, is insufficient to maintain the blood calcium level, it is withdrawn from the fairly large reserves in the bone. Only when the bone reserves are depleted does the blood calcium level fall, causing tetany (muscular twitching) which can be fatal. A temporary alteration of the blood calcium and tetany can be induced by over breathing. It is alleviated by rebreathing into a paper bag. In normal health, that is when the skeleton has not been depleted, calcium in food is not a 'sedative'.

In practice, the skeleton is rarely depleted by a shortage of calcium alone in the diet—most diets supply at least 200 milligrams per day, which is usually sufficient to maintain the reserves. It can be depleted, however, by an insufficiency of vitamin D, as in old age; by increased needs, not compensated by an increase in calcium and vitamin D (for example in pregnancy); and if nutrients are not absorbed into the blood stream as a result of digestive disorders (see Malabsorption). In rare cases disease of the parathormone-producing glands can cause too much hormone to be secreted, depleting the bones of calcium and causing kidney stones. It is cured by surgical removal of part of the over-active glands.

In pregnancy and breast feeding, calcium is supplied to the growing child, regardless of the adequacy of the mother's diet, by withdrawal from the skeleton. A low intake of vitamin D and calcium, coupled with successive pregnancies results in gradual weakening of the bones—osteomalacia. Osteomalacia frequently occurs in old age, particularly in women.

For the majority, a high calcium diet is not harmful because intakes in excess of needs are not absorbed into the blood stream, but there are some people who absorb more calcium than normal from food. See below. Vitamin D is toxic when taken in excess: too much calcium is absorbed from food and deposited in blood vessels, with sometimes fatal results.

Low calcium diets may be prescribed for the treatment of idiopathic hypercalcuria, a condition in which the proportion of calcium absorbed from food is greater than normal. As a result, abnormally large quantities of calcium are eliminated from the body in urine. Unless the urine is kept dilute, deposits of calcium tend to form in the kidney (kidney stones) and may eventually obstruct the flow of urine, resulting in kidney failure. Small stones may be passed (and cause much pain). Larger stones may have to be surgically removed. Idiopathic hypercalcuria can be treated by a low calcium and high water diet but this treatment may not be suitable for the many other causes of kidney stones. At least 4 pints of water or other beverages should be drunk. More may be needed in hot climates or working conditions. Usually, avoidance of milk, cheese and yoghurt is sufficient: other foods may have to be avoided in more serious cases.

CALORIE—other name kilocalorie (kcal)

An expression of the **energy** contained in food and required for daily life and activity. It is a unit of heat—one of the many forms of energy. One calorie is the amount of heat required to raise the temperature of 1 gram of water by 1°C.

The calorie is a very small unit, consequently the Calorie, which is 1000 times greater than the calorie, has always been used for nutritional measurements. The kilocalorie is synonymous with Calorie and avoids confusion, and resultant printing errors, between calorie and Calorie.

Kilocalories are soon to be replaced by the general unit for measuring energy, the kilo**joule**. However, since most lay people are unfamiliar with this unit, kilocalories have been retained in the text entries, apart from energy.

CARBOHYDRATE—chemical names, mono, di and polysaccharides

In human nutrition, carbohydrate refers to **starches** and **sugars**. **Cellulose**—the most abundant carbohydrate in nature—and other undigested **polysaccharides** contribute to the dietary **fibre**.

Sugar (sucrose), glucose powder and refined starches are the most concentrated source of carbohydrate. Flour, cakes, bread, biscuits, breakfast cereals, pastry, honey, potatoes and preserves are also good sources. Fruits, vegetables (especially pulses), nuts, alcoholic drinks and milk contain small amounts. Fats, cheese, eggs, meat and fish contain none. The following table shows quantities of foods containing 10 grams of carbohydrate.

Carbohydrates are used in the body for **energy**: starches and **dextrins** are split to their component units of glucose during **digestion**; sucrose to its component units of **glucose** and **fructose**; lactose (milk sugar) to its component units of glucose and **galactose**. These simple sugars are absorbed into the blood stream and used either immediately for energy or stored for later use in the form of **fat** or **glycogen**.

All carbohydrates contain the same amount of energy—approximately 4 kilo**calories** per gram. Glucose and sugar are not superior sources of energy, although they may be absorbed into the blood stream at a faster rate.

There is no recommended intake for carbohydrate in the diet because fats, proteins and carbohydrates all supply energy to the body. The average British daily diet contains about 300 grams (excluding soft drinks and sweets taken outside the home), which supplies about half the total energy in the diet. About half of the intake of carbohydrate (140 grams) is derived from starch in cereals, flour, bread and potatoes.

Nearly all foods are mixtures of **nutrients** and correspondingly most carbohydrate containing foods are valuable sources of essential nutrients in normal diets—notably B

vitamins and **protein** in bread; and **vitamin C** in potatoes. Refined starches and sugars are an exception: they are virtually pure carbohydrate ('empty calorie' foods) and can be eliminated from the diet with no risk to health.

Further restriction of carbohydrate—to about 100 to 150 grams per day or 10 to 15 ten gram portions—(a low carbohydrate diet) is perhaps the most convenient method of reducing weight. Provided the diet contains sufficient meat, cheese and vegetables, a **balanced diet**—containing adequate amounts of protein, vitamins and minerals—will be ensured. Total elimination of carbohydrate is inadvisable, except under medical supervision—see slimming diets.

Controlled intakes of carbohydrates are prescribed for **diabetes** and some disorders of blood **lipids**.

QUANTITIES OF FOODS CONTAINING APPROXIMATELY
10 GRAMS CARBOHYDRATE

	Food	Portion containing 10 grams sugar or starch
Sugar	White or brown	10 grams (2 level teaspoons)
	Boiled sweets	10 grams (⅓ oz)
	Jam, marmalade or honey	15 grams (1 teaspoon)
	Toffees	15 grams (½ oz)
	Chocolate	20 grams (⅔ oz)
	Others	a
Cereals	Flour	15 grams (1 level tablespoon)
	Tapioca, sago, rice, cornflour, custard powder, other thickenings	10 grams (2 heaped teaspoons)
	Unsweetened breakfast cereals	15 grams (3 heaped tablespoons)
	Boiled rice	35 grams (1 tablespoon)
	Boiled spaghetti, macaroni	40 grams (2 tablespoons)
	Bread (all)	20 grams (½ large slice)
	Crispbread	15 grams (½ oz)
	Cream crackers	20 grams (2 biscuits)
	Sweet biscuits	15 grams (2 'marie' type)
	Shortcrust pastry and scones	20 grams (⅔ oz)
	Other cakes and biscuits	a
Fruit	Currants, dates, sultanas, raisins	15 grams (½ oz)
	Bananas (without skin)	50 grams (½ large)
	Grapes and tinned fruit	60 grams (2 oz)
	Raspberries, strawberries	180 grams (6 oz)
	Most other fresh fruit	120 grams (1 medium apple etc)
Vegetables	Lentils and other dried pulses	20 grams (⅔ oz)
	Chips	30 grams (about 4 large)
	Boiled potatoes	50 grams (2 oz)
	Baked beans, sweetcorn	50 grams (2 oz)

	Food	Portion containing 10 grams sugar or starch
Vegetables	Parsnips (boiled)	75 grams (2½ oz)
	Beetroot (boiled)	100 grams (about 1 medium)
	Peas, broad beans, fresh or frozen	130 grams (4 tablespoons)
	Carrots (boiled)	230 grams (very large portion)
	Most other vegetables are virtually carbohydrate free	
Dairy foods	Milk, fresh	200 grams (1 glass)
	Milk, dried skimmed	20 grams (5 heaped teaspoons)
	Milk, evaporated	80 grams (2 tablespoons)
	Yoghurt, low fat unsweetened	150 grams (1 carton)
	Single cream	300 grams (½ pint)
	Cottage cheese	220 grams (8 oz)
	Other cheese, butter and double cream is virtually carbohydrate free	
Meat, fish, eggs	Sausages, cooked	70 grams (2½ oz)
	Other made up meat and fish foods	a
	Fresh meat, fish (except some shell fish) and eggs are carbohydrate free	
Beverages	Port	90 mls (a small glass)
	Wine, sweet (Sauternes)	170 mls (1 glass)
	Beer (bitter)	330 mls (just over ½ pint)
	Soft drinks	a
Miscellaneous	Ice cream	50 grams (1 small brick)
	Peanuts (shelled)	120 grams (4 oz)
	Chestnuts (shelled)	25 grams (1 oz)
	Soups (clear are virtually carbohydrate free)	a
	Fats and oils are carbohydrate free	

a – made up and manufactured products are very variable. Booklets containing information about brand products are available from some supermarkets

CAROTENE—also called provitamin A

A precursor of **vitamin A**. It is an orange pigment, found in yellow and orange fruits and vegetables. Green vegetables also contain carotene, masked by the green pigment, chlorophyll. White vegetables and fruits contain none.

Carotene contents are proportional to the orange or yellowness of fruits and vegetables: old carrots contain more than young ones. In some red and purple fruits and vegetables, the colour is due to anthocyanins, which do not have vitamin A activity. In green vegetables, the carotene content is proportional to greenness: outer dark leaves of lettuce and cabbage contain much more carotene than the inner pale ones.

About one third of the total vitamin A in the average diet is supplied by carotene. When butter and margarine were rationed during the war, carotene supplied half the vitamin A equivalents.

Several forms of carotene exist. The most common and potent form, β carotene, is referred to here. α carotene (in palm oil), and cryptoxanthene, in maize, have only half the potency of β carotene.

During **digestion**, carotene is converted to vitamin A in the cells lining the small intestine, and together with other vitamin A in food, transported to the liver. In vegetables and fruits only one sixth is converted to vitamin A. This is partly because carotene is poorly absorbed from fat free foods, but there are other unexplained factors. The availability of carotene is taken into account when assessing the vitamin A values of foods. The table below shows carotene (expressed as vitamin A equivalents) in different fruits and vegetables.

Animals and fish convert carotene in grass and feeds to vitamin A, which is colourless. In cows, some carotene escapes conversion and is passed into the milk, together with vitamin A. The creamy colour of milk fat (in milk, butter,

CAROTENE IN FRUITS AND VEGETABLES

	Fruit or vegetable	Average vitamin A (microgram equivalents) in a 100 gram (3½ oz) portion
Rich	Carrots—old	2000
	—canned	1200
	—new	1000
	Spinach and turnip tops	1000
	Kale, beetgreens	800
	Watercress	500
Good	Broccoli tops	400
	Endive, yellow melon (none in white)	300
	Pumpkin, raw apricots	250
	Stewed dried apricots	200
	Canned apricots	170
	Tomatoes, fresh	120
	Raw peaches, canned tomatoes, stewed prunes, green beans	80

All other yellow/orange/green fruits and vegetables contain carotene but generally in small amounts (less than 80 equivalents). White and some dark red fruits and vegetables like beetroot, red cabbage) contain none.

cream, etc.) is due to carotene, but in eggs, the yellow colour of the yolk is due to another pigment which does not have vitamin activity. Carotene is added to vegetable (kosher) margarine for both colouring and enrichment purposes. In fatty foods—margarine and dairy products—3 times as much carotene is absorbed, and therefore half the carotene is converted to vitamin A.

In humans, carotene can also escape conversion to vitamin A, and is stored in the layers of fat under the skin. Great excesses of carotene, for example from a diet consisting only of carrot juice and oranges, can cause a yellow skin. In contrast to vitamin A intoxication, this condition, carotenaemia, is thought to be harmless.

There are little or no losses of carotene in cooking because it does not dissolve in water and is stable to boiling. Canned and frozen vegetables retain almost all of their carotene. However, carotene is destroyed (oxidised) in the presence of oxygen in air and ultra-violet light in daylight. **Antioxidants** in fats and vitamin C and the **preservative** sulphur dioxide in fruits and vegetables protect against these losses. Sun-dried pulses and currants contain no carotene, but apricots and peaches that have retained their colour will contain some.

AVERAGE NUTRIENTS IN 20 GRAMS DRY IRISH MOSS
(ENOUGH TO MAKE ONE PORTION JELLY)

Energy (kilojoules)	25
(kilocalories)	5
Protein (grams)	1½
Carbohydrate, fat	Trace
Sodium (milligrams)	580
Potassium (milligrams)	420
Calcium (milligrams)	170
Magnesium (milligrams)	125
Iron (milligrams)	1·8
Phosphorus (milligrams)	40
Other minerals	Rich in iodine
Vitamins	—(may contain vitamin B_{12})

— – no figures available

There are smaller losses of carotene in modern methods of dehydration. Carrots lose about 40% when dried by hot air and about 20% when freeze dried.

CARRAGHEENIN

A red seaweed, which dissolves when boiled and sets to a firm jelly. The jelly is usually called carragheen or Irish Moss and can be eaten as a sweet. It is a rich source of minerals, especially **iodine**, but may have to be avoided in **galactose**-free diets (it contains a **polysaccharide** of galactose which, although not normally digested, may be partially split by bacteria in the large bowel). Laver, another red seaweed also rich in iodine, is used for laver bread.

Carragheen is sometimes used as a stabiliser and thickener (see Emulsifiers) by the food industry.

CELLULOSE

A component of dietary **fibre** found in all plants. It is a **polysaccharide** composed of glucose units but humans possess no digestive **enzymes** capable of breaking the linkages apart. Tough raw vegetables are not well digested. Cooking softens the cellulose cell walls and renders the nutrients inside more accessible for digestion.

Methyl cellulose is the basis for some slimming products. It swells in the stomach and is said to reduce appetite. It may also be used for proprietary constipation remedies: **Bran** is less expensive and as effective.

Bacteria that inhabit the digestive systems of ruminants supply the necessary enzymes for dismantling cellulose. Hence cows and sheep are able to digest grass and use the liberated energy.

CEREALS

Strictly the seeds of cultivated plants in the grass family. The seven major cereals grown are barley, maize, millet, oats rice,

rye and wheat—see individual entries. The term 'corn' refers
to the indigenous cereal—wheat in England and maize in
some parts of America.

Cereals have a similar nutritional value (see table), containing mostly **starch**. Although wholegrain cereals contain
small quantities of all other **nutrients** they cannot support
life when eaten alone because they are lacking in vitamins A
(except yellow maize), B_{12} and C. Wholegrain cereals also
contain **phytic acid** which interferes with the absorption of
iron, calcium and some trace elements. They are however a
valuable base (staple), supplying **energy, protein** and **vitamins
B,** in the diet which can be balanced with small quantities of
vegetables and animal foods (see Balanced diet).

About 2% of the weight of cereals is the prospective plant
(germ), about 13% the outer husk (bran) and about 85% the
inner starchy food store (endosperm). Most of the B vitamins, fat, **fibre,** iron, trace elements and vitamin E are concentrated in the bran, germ and outer layers of the endosperm. Protein is distributed throughout the wholegrain.

The bran and germ are easily removed from rice and
wheat. Highly milled (refined) cereals are more palatable
(because they are lacking in fibre) and easier to store (because they are lacking in fat) but of comparatively poor
nutritional value. Eighty per cent, for instance, of vitamin
B_1 is removed when rice is milled and polished. Some of the
nutrients are partly replaced in white **flour** in Britain, but
white flour is still a comparatively poor source of fibre,
vitamin E and B_6 (pyridoxine) and some trace elements
compared to wholemeal flour.

All cereals can be ground to flour but in Western countries
wheat is the preferred cereal because it contains sufficient
gluten to make acceptable leavened bread. The other cereals
are chiefly used for animal feed, brewing and breakfast
cereals. Breakfast cereals are toasted or puffed, which may
destroy the B vitamins—in particular vitamin B_1. Some
manufacturers replace these losses—see cereal entries.

Other starchy dry foods eaten in Britain—sago, tapioca

and arrowroot—are sometimes called cereals. They are plant food stores but not seeds and consequently do not contain the germ and are low in protein. They are of very inferior nutritional value, but fortunately are not staple foods in Britain and of little importance in the diet.

Starchy foods are the cheapest article of the diet and may supply at least 70% of the total energy in impoverished communities. Their nutritive value is therefore of crucial importance in determining the health of consumers. The lack

AVERAGE NUTRIENTS IN 100 GRAMS (3½ OZ) WHOLEGRAIN CEREALS

Nutrient	Barley	Maize	Millets	Oats	Rice	Rye	Wheat	Cassava[a]
Energy								
(kilojoules)	1425	1525	1490	1630	1405	1470	1445	1435
(kilocalories)	340	365	355	390	355	350	345	340
Protein (grams)	12	10	10	12	8	8	8–15	1·5
Fat (grams)	2	4·5	2·5	7·5	1·5	1·5	2	Trace
Carbohydrate (grams)	70	70	70	70	80	80	70	85
Calcium (milligrams)	35	12	20	60	10	25	30	55
Iron (milligrams)	4	2·5	5	5	2	3·5	3·5	2
Vitamins A and C	0	0	0	0	0	0	0	0
B vitamins, thiamin (milligrams)	0·5	0·35	0·6	0·5	0·25	0·3	0·4	0·04
Other B vitamins ⎫ Vitamin E ⎬ See individual cereal entries Other minerals ⎭								

a – Cassava (tapioca) is not a true cereal

of protein in cassava (tapioca) is the main cause of the disease **kwashiorkor** in communities subsisting on it. **Pellagra** used to be endemic in the southern states of America, where maize was the main food in the diets of the poor. **Beriberi** has not yet been completely eradicated from some communities where polished rice is the staple cereal. When circumstances —usually income—allow a variety of foods to be eaten the diet is more easily balanced and the nutritive value of the staple cereal is of less importance. However, a lack of **fibre** and of **trace elements** are possible factors in several degenerative diseases in Western countries—see also Heart disease.

CHEESE

Most cheeses are excellent sources of **protein, calcium** and **phosphorus**. They are good sources of **riboflavin**, and supply other B **vitamins**, but are lacking in vitamin C. All except those made with skim **milk** (like some cottage cheeses) are good sources of **vitamin A**, and contain vitamins D and E. All cheeses are poor in iron, but supply other **minerals**.

Although it is virtually **carbohydrate** free, most cheese is high in **fat** and therefore **energy**. Weight for weight, cheddar contains about 3 times more energy than bread. It also contains **cholesterol** and a high proportion of saturated **fatty acids**. It should be eaten with care in **slimming diets**: some cottage and curd cheeses made only with skim milk are much lower in energy and can be eaten liberally. Cream cheese is the highest in energy (see table below), and cheeses made with added cream, like Stilton, are also high.

All cheeses, except cream cheese, are sources of protein. Cheddar contains about a quarter of its weight as protein. Cheddar and Parmesan are the best sources of calcium: about 60 grams (2 oz) of cheddar will supply the adult **recommended intake** for calcium. Soft cheeses, like cottage, tend to be lower in calcium than other cheeses—they are made in a similar way to **yoghurt** and the calcium combines with the acid and is carried out in the whey.

Processed cheeses are pasteurised (see milk) to prevent overripening, treated with emulsifiers to prevent separation of the fat and packed into moisture-proof coverings. Processed cheeses have a slightly lower protein and higher water content than ordinary cheese. Cheese spreads are treated in approximately the same way as processed cheese, but they are appreciably higher in water and lower in nutritional value than ordinary cheese—and more expensive.

Hard cheese is permitted to contain the **preservatives** nisin and sorbic acid and specified permitted natural **colours** or their synthetic equivalents. Mineral hydrocarbons are permitted on the rind. Additionally soft cheese, cheese spreads,

AVERAGE NUTRIENTS IN 25 GRAMS (1 OZ) OF SOME CHEESES

Nutrient	Camembert	Cheddar	Cheshire	Cream	Edam	Gorgonzola	Parmesan	Processed	Spread	Stilton	Cottage
Energy (kilojoules)	325	445	410	380	330	415	440	395	305	500	125
(kilocalories)	75	105	95	90	80	100	105	95	75	120	30
Protein (grams)	6	6½	6½	¾	6	6¼	8¾	5¾	4½	6½	4
Fat (grams)	6	8¾	7½	10	5¾	7¾	7½	7½	5¾	10	1[a]
Cholesterol (milligrams)	20	25	—	25	25	—	25	20	—	—	3·5
Carbohydrate (grams)	0	0	0	0	0	0	0	0	0	0	1
Calcium (milligrams)	40	200	155	7	185	135	305	180	125	90	20
Phosphorus (milligrams)	70	135	115	10	130	95	195	120	110	75	—
Sodium (milligrams)	350	155	175	30	245	305	190	230	290	290	80
Potassium (milligrams)	30	30	25	10	40	45	40	20	40	40	—
Magnesium (milligrams)	5	10	5	1	5	10	10	10	5	5	—
Other minerals	Cheeses supply other minerals and trace elements; they are low in iron.										
B vitamins:											
riboflavin (milligrams)	0·2	0·13	0·13	0·05	0·1	—	—	0·1	—	0·08	0·07
nicotinic acid (milligram equivalents)	1·6	1·5	—	—	—	—	—	1·4	—	—	1
vitamin B_{12} (micrograms)	—	0·2	0·2	—	—	—	—	0·2	—	—	—
others	Cheeses supply small amounts of other B vitamins										
Vitamin E (α tocopherol milligrams)	—	0·25	—	—	—	—	—	—	—	—	—
Vitamin A (microgram equivalents)	60	105	90	120	70	90	90	90	70	120	7
Vitamin D (micrograms)	0·05	0·09	0·08	0·2	0·06	0·08	0·08	0·08	0·06	0·1	0·005
Vitamin C	0	0	0	0	0	0	0	0	0	0	0

— – no figures available

a – some cottage cheese, made from skim milk, is fat free

and processed cheese may contain **flavourings** and certain specified **emulsifiers** (like gums, alginates and lecithin). All except cheddar, cheshire and soft cheese may contain **nitrite**.

The national average intake of cheese is low—about 3¾ oz per head a week—and, despite its nutritional value, it contributes little to the total average intake of nutrients. It has an unfounded reputation for being indigestible—probably because its high **fat** content prolongs the time cheese remains in the stomach. It hardens and becomes less digestible when cooked but this can be prevented if it is combined with starch to prevent separation of the fat. When grilling cheese on toast, it is advisable to mix the grated cheese with a little milk, cornflour or mustard and cook under a hot grill for as short a time as possible.

Cheese and drugs

Tyramine (see Tyrosine) occurs naturally in cheese, extracts, baked beans, alcohol and yoghurt. It is normally detoxicated by a group of enzymes called the monoamine oxidases (MAO), but certain anti-depressant drugs, such as Nardil, inhibit their action. These foods must not be eaten in conjunction with MAO inhibiting drugs: an alarming rise in blood pressure—with sometimes fatal results—has been recorded.

CHLORINE

An **electrolyte**, closely associated with **sodium** in the body and food. Adults contain about 70 grams of chlorine, most of which is in the fluids surrounding cells.

The majority of chlorine in the diet is derived from salt (sodium chloride). Salt-free diets contain about ½ gram of chlorine; diets containing salt in cooking and added as a preservative contain 5–9 grams per day. Chlorinated water supplies a negligible amount—about 3 milligrams per day.

Chlorine is absorbed out of food and discarded from the body in urine and sweat with sodium. Although it is an essential nutrient, deficiency of chlorine does not occur in otherwise normal health without a corresponding deficit of sodium.

CHOCOLATE AND CONFECTIONERY

By weight, chocolate and confectionery is at least half sugar (**sucrose**) which is unsuitable for diabetics, weight reducers, and known to be harmful to **teeth**. Chocolate, made from **cocoa**, sugar, cocoa butter and flavourings, contains up to 40% fat—mostly saturated—and should be avoided in **cholesterol** lowering diets. Diabetic chocolate is not suitable for weight reducers—sugar is replaced by **sorbitol**, which has the same energy content.

Pastilles, fruit gums, chocolate and liquorice contain

AVERAGE NUTRIENTS IN 25 GRAMS (1 OZ) CONFECTIONERY

Nutrient	Boiled sweets and peppermints	Plain chocolate	Milk chocolate	Fruit gums	Liquorice Allsorts	Pastilles	Toffees
Energy (kilojoules)	375	570	615	180	330	265	455
(kilocalories)	90	135	145	45	80	65	110
Protein (grams)	Trace	1½	2	¼	1	1¼	½
Carbohydrate (grams)	25	15	15	10	20	15	20
Fat (grams)	Trace	9	9	0	½	0	4½
Calcium (milligrams)	Trace	15	60	90	10	10	25
Iron (milligrams)	0·08	0·7	0·4	1	2	0·3	0·4
Potassium (milligrams)	Trace	65	85	90	55	10	50
Sodium (milligrams)	10	35a	70a	15	20	20	80a
Magnesium (milligrams)	Trace	35	15	25	10	3	5
Vitamin A (microgram equivalents)	0	1·5	1·5	0	0	0	0
B vitamins:							
thiamin (milligrams)	0	0·008	0·008	0	0	0	0
riboflavin (milligrams)	0	0·09	0·09	0	0	0	0
nicotinic acid (milligram equivalents)	0	0·6	0·6	0	0	0	0

a   contains added salt: these weets may have to be avoided in low *sodium* diets minerals and small amounts of protein (milk chocolate has more calcium and protein, from milk, than plain) but boiled sweets and peppermints are sources of 'empty calories'. Toffee and fudge contain added salt and may have to be avoided in low **sodium** diets.

Confectionery is exempt from the need to declare ingredients on labels, but most food **additives**—including flavours, colours, mineral hydrocarbons—are permitted. Antioxidants and preservatives are permitted in ingredients and may be 'carried over' into the finished food.

CHOLESTEROL

Cholesterol is part of the body structure and a normal constituent of the blood stream. Most adults contain about 140 grams (5 oz). It is needed for cell membranes—particularly nerves—and also for the synthesis of some hormones, bile salts, and vitamin D. Animal foods contain cholesterol, but it is not an essential **nutrient**: the body normally synthesises at least twice as much as is eaten in food each day.

Foods containing the most cholesterol are brains and egg yolk. Other foods contain much smaller amounts (see Table 1) and plant foods none. It is interesting that lean meats do not contain much less cholesterol than fatty ones: the cholesterol in lean beef and dripping for example is about the same.

Daily cholesterol intakes vary—mainly depending on the number of eggs consumed—from 0 to 3000 milligrams. A possible intake from 1 egg, 2 portions of meat or fish, and 1 oz butter is about 500 milligrams, of which about half is absorbed into the blood stream. An extra egg will add a further 300 milligrams.

Normally cholesterol gains—from body synthesis and diet—are balanced by losses, mainly in the bile. However, people living in Western societies tend to accumulate cholesterol, and as a result, the amount circulating in the blood stream often rises with increasing age.

TABLE 1

APPROXIMATE CHOLESTEROL CONTENTS OF SOME FOODS (1)

Cholesterol rich foods	more than 1 gram (1000 milligrams) in 100 grams food: Brains, egg yolk; 200 to 600 milligrams in 100 grams food: Butter, roe (including caviar), sweetbreads, kidney, liver (all); 100 to 200 milligrams in 100 grams food: Most cheese, crab, double cream, haddock, heart, sardines, shrimps
Moderate in cholesterol	50 to 100 milligrams in 100 grams food: Beef, beef fat, chicken, chicken fat, cod, single cream, frogs' legs, herring, lamb, lamb fat, lobster, mackerel, oysters, pork, pork fat, rabbit, mayonnaise, tuna, trout, turkey, veal, veal fat
Low in cholesterol	less than 50 milligrams in 100 grams food: Cottage cheese, ice cream, margarine with added butter, milk (skim less than full cream), salmon, yoghurt
Cholesterol free foods	Egg white, all plant foods, like oils, fruit, nuts, vegetables, cereals, cooking fats and margarine made entirely from plant oils and fats, jams, sweets

(1) taken from Feely et al. (1972) *J. Am. Dietet. A.* 61 134

A raised blood cholesterol (above 150 milligrams per 100 millilitres of plasma) is a main risk factor in **heart disease**: when combined with another risk factor, like cigarette smoking, the likelihood of dying from heart disease is increased. Blood cholesterol can be lowered by eating less cholesterol and, more effectively, by cutting down on saturated **fatty acids** in animal **fats** (particularly beef fat, dairy products and lamb). Polyunsaturated fatty acids found in most plant **oils** also decrease the blood cholesterol— probably by hastening the loss of cholesterol in bile. However, they are less effective in lowering blood cholesterol than saturated fats are in raising it, and should not be used in addition, although—provided there is no overweight they can be used to replace saturated fats. See also Fibre, Pyridoxine, Trace elements, Sucrose.

Blood cholesterol lowering diets (low in cholesterol and saturated fats, and high in polyunsaturated fats) are usually necessary for some people with disorders of blood **lipids**. An example is shown in Table 2.

TABLE 2

AN EXAMPLE OF A BLOOD CHOLESTEROL LOWERING DIET

Avoid	Butter, hard *margarine*, other animal and hydrogenated *cooking fats*, fatty meats
	Ordinary (full cream) milk, cream, cream cheese
	Bought cakes, biscuits, pastries
	Chocolate, cashew nuts, coconut, ice cream, mayonnaise
Limit	To two portions per week, foods rich in cholesterol (see table 1)[a]—for example allow one egg and one portion liver per week
Take moderate portions of	Lean meat, fatty fish, white fish, olive oil, olives, avocado pears
Eat liberally	Vegetables, salads, fruits, potatoes, cereals, other nuts, cornflour and other thickenings
	Skim milk (fresh or dried), fat-free cottage cheese and yoghurt, egg white
	High polyunsaturated margarines and oils, like Flora, corn (maize) oil, safflower oil, sunflower oil, soya oil, peanut (groundnut or arachis) oil.

a – these foods may be forbidden in very strict low cholesterol diets

TABLE 2: SAMPLE MENU

Breakfast	Fruit juice
	Cereal (preferably wholegrain, like porridge or wheat flakes) with skim milk and a little sugar, and/or
	Lean bacon or fish with bread (preferably wholemeal) fried in permitted oil if wished
	Bread or toast (preferably wholemeal) with polyunsaturated margarine and little marmalade or jam
Lunch	Sandwiches (for instance tuna fish and tomato) made with polyunsaturated margarine and preferably wholemeal bread
	Fresh fruit, or low fat yoghurt with chopped nuts and raisins if wished
Supper	Melon, fruit juice, clear soup, or thick soup made with flour, skimmed milk, polyunsaturated margarine, and fat-free stock
	Any low or moderate in cholesterol lean meat or fish, for example chicken, in for instance pie (pastry made with flour, and permitted oil or polyunsaturated margarine)
	Any vegetable, rice or pasta boiled or baked or fried in permitted oil, or with polyunsaturated margarine, for instance baked carrots and jacket potato
	Fruit or milk pudding made with cereal, skimmed milk and little sugar
Drinks	Tea or coffee with skim milk and preferably no sugar
	Soft drinks
	Alcohol, if permitted
	Avoid cocoa and malted milks

This type of diet may also lessen the likelihood of heart disease for people who are unable to give up smoking, or who have another risk factor, such as high blood pressure. However, the members of a recent Government Committee concluded that such radical alteration of the diet is probably of limited advantage for most people. Stopping smoking, avoiding overweight and taking plenty of regular exercise are more important general measures. Although blood cholesterol lowering diets used in several trials, including American, apparently reduced the risk of heart disease, their effect may not have been due to polyunsaturated fats alone, but also to improved general nutrition. These diets contain good quantities of fibre, trace elements and vitamins.

CHOLINE

In diets containing sufficient **protein**, choline can be made in the human body, notably from **methionine**, an **amino acid**. Most people probably synthesise about 10 milligrams a day. Although it is not an essential **nutrient** for man (in contrast to some animals) choline is abundant in food. For example one egg and 100 grams ($3\frac{1}{2}$ oz) of meat and cereal contain 500, 600 and 100 milligrams respectively.

Choline has several important functions in the body, including the formation of lecithin and other **phospholipids**. When the diet is low in protein, for instance in alcoholism and **kwashiorkor**, insufficient choline may be formed. This may cause accumulation of fat in the liver.

CHROMIUM

A **trace element**, only recently recognised as an essential **nutrient** for humans. It enhances the action of insulin—the main hormone controlling the utilisation of **glucose** in the body (see Diabetes)—but this is probably not its only function in the body.

During **digestion**, only small amounts of chromium are

absorbed from food into the blood stream. One of the several forms of chromium (called Glucose Tolerance Factor) is probably absorbed more efficiently and may be more potent than other forms. As yet, estimates of either total chromium or GTF contents of foods are uncertain. The best source of GTF found so far is brewer's yeast. Diets high in fats, refined starches and sugar contain less chromium than those including plenty of fresh vegetables and whole-grain cereals. Large amounts of chromium (which is labile to heat) may be lost during cooking and processing.

The daily needs for chromium are thought to be 5–10 micrograms (one microgram is one millionth of a gram). Depending on the source, 20 to 500 micrograms might be needed in food to replace daily losses. American diets reportedly contain 5 to 100 micrograms per day.

Needs for chromium are increased when the diet is excessively high in sugar or starch. These nutrients require more insulin for their utilisation and correspondingly induce more chromium to be released from tissues and lost in the urine after exerting its effect. Since refined starches and sugars are poor sources of chromium, even moderate intakes may cause a gradual depletion of body stores, resulting in mild deficiency in middle or old age. The body content of chromium is reported to fall with increasing age in Americans, but there have been no comparable studies in the UK.

Under certain circumstances, severe chromium deficiency in animals causes a syndrome similar to diabetes. Whilst chromium supplements cannot cure frank diabetes in humans, mild intolerance to **carbohydrates** (resulting in occasional glucose losses in urine—see Diabetes) has been improved in middle-aged and elderly people deficient in chromium. Chromium deficiency is also likely to occur in diabetics treated with insulin injections (which are thought to increase chromium losses in the urine), after successive pregnancies, and prolonged subsistence on formula diets deficient in chromium. It frequently occurs in children suffering from the **protein** deficiency disease **kwashiorkor.**

Other symptoms occuring in animals fed chromium deficient diets are raised blood **cholesterol**, increased incidence of **atherosclerosis** and decreased life span. The possible preventative effect of chromium in heart disease—in view of suspected large losses in food processing—is under investigation.

CITRIC ACID

Widely distributed in fruits (especially citrus fruits), vegetables and yoghurt. Commercially manufactured and used as an acid flavouring agent in jams, soft drinks and sweets.

Citric acid is continually made and broken down in the body. It takes part in a vital chain of reactions called the citric acid (or Kreb's) cycle whereby **energy** is liberated from food. See glucose and metabolism. Carbon dioxide and hydrogen (which is later combined with oxygen to form water) are the end products: they are removed from the blood stream by the lungs and kidneys.

COBALT

Cobalt occurs in its free form in plants, but it is thought to be utilised by man only as part of the essential nutrient **vitamin B$_{12}$**, supplied in animal foods. Like other **trace elements**, cobalt is toxic in excess. Doses 1000 times greater than normally present in food have caused heart failure. It is more toxic when taken with alcohol: before this was realised, cobalt salts were used as foaming agents in beer.

Work with animals suggests that cobalt may be an antagonist to another trace element, **iodine**.

COCOA

A beverage containing **protein, minerals** and small quantities of B vitamins and vitamin A. It is a rich source of iron, but probably little is available for absorption into the blood stream. Because so little is used, cocoa is not an intrinsically important source of **nutrients**, but it is a nourishing drink when made with milk.

Cocoa contains the stimulants **caffeine** (about 20 milligrams per cup) and theobromine (about 200 milligrams per cup). Theobromine is a weaker stimulant than caffeine.

Cocoa nibs (the separated beans) contain up to 60% of their weight in **fat** (cocoa butter) rich in saturated **fatty acids**. It is partially removed before the nibs are ground: otherwise it would separate out from the beverage. Although cocoa powder contains less saturated fat (20 to 25%) than nibs, it should be used in moderation in **cholesterol** lowering diets. Darker powders contain more fat and have a richer flavour.

Drinking chocolate contains cocoa (usually treated with alkali to prevent less sediment forming in the beverage) and sugar. It contains about twice as much carbohydrate and less protein, iron and minerals than cocoa. Milk powder, salt and vanilla flavour may also be added.

COMPARISON OF 15 GRAMS (3 TEASPOONS) DRINKING CHOCOLATE
AND COCOA POWDER

Nutrient	Cocoa	Drinking chocolate
Energy (kilojoules)	285	260
(kilocalories)	70	60
Carbohydrate (grams)	7	13
Protein (grams)	3	1
Fat (grams)	$3\frac{1}{2}$	1
Iron (milligrams)	2·3	1·8
Vitamin A (microgram equivalents)	1	0·3
Thiamin (milligrams)	0·01	0·005
Riboflavin (milligrams)	0·05	0·02
Nicotinic acid (milligram equivalents)	0·7	0·2
Other B vitamins	—	—
Vitamin C	0	0

— – no analyses available

COELIAC DISEASE

A common form of **malabsorption**, causing loss of weight and vitamin and mineral deficiencies. It is estimated to affect about one in every 2000 people in Britain.

Coeliacs are sensitive to gliadin, part of the protein **gluten** in wheat, rye, and to a lesser extent barley and oats. Most are affected in childhood after gluten cereals are first introduced into the diet. Perhaps because of the absence of an **enzyme** necessary for its **digestion**, gliadin irritates the cells lining the villi in the small bowel, which are responsible for absorbing nutrients and transferring them to the blood stream. The villi become inflamed and eventually degenerate, when nutrients are not absorbed, but are carried out of the body, usually as fatty diarrhoea. The child fails to thrive, is often irritable, has a characteristic pot belly, and may suffer from **anaemia** and **rickets**. Milder forms may pass unnoticed, but in adult life may result in general ill health, loss of weight, tiredness and **osteomalacia**.

Severe coeliac disease used to be fatal, but now is cured by complete removal of gluten from the diet. All untreated wheat flour must be excluded and barley and rye are usually prohibited. Treatment in hospital is usually necessary to allow regeneration of the villi and full recovery.

Maize (for instance in cornflakes, cornflour), rice and potatoes are gluten free. These starches, together with fresh meat, fish, cheese, eggs, vegetables and preserves are allowed freely. Special gluten free biscuits, pasta, bread and flour (which can be made into bread, cakes and biscuits) made from treated wheat are available on prescription. Many convenience foods are unsuspected sources of gluten (for example pepper compound, vending machine coffee) and to enable successful adherence to the diet, instruction should be given by a qualified dietitian. The Coeliac Society, P.O. Box 181, London NW2 2QY, publishes an annual list of gluten free convenience foods and is a valuable source of help and information.

Some children are apparently able to tolerate gluten eventually but, because anaemia, osteomalacia and more serious complications can develop insidiously and without discomfort, it is usually necessary for children to adhere to a lifelong gluten-free diet. Coeliacs diagnosed in adulthood are rarely able to return to a normal diet.

COFFEE

A strong cup of coffee, made from 2 oz of coffee in one pint of water, contains approximately 150 milligrams of **caffeine**, 150 milligrams of **potassium**, and 2 milligram equivalents of **nicotinic acid**. Black coffee contains hardly any **energy**.

Roasted chicory root (which has a bitter flavour but is lacking in caffeine) is added to French coffee. Legally French coffee must contain a minimum of 51 % by weight of coffee, and Viennese coffee (coffee with figs) 85 %.

Instant coffees are spray- or freeze-dried coffee infusions. One teaspoon (2 grams) of instant coffee contains about 100, 1, and 50 to 120 milligrams potassium, nicotinic acid, and caffeine respectively.

Coffee essences contain concentrated coffee infusions and sugar. Both instant and coffee essence may contain **preservative** and **emulsifier**. Legally, decaffeinated coffee must not contain more than 0·1 % caffeine (approximately 2 milligrams per cup).

COLOUR

Colours are added to food to replace losses (peas for example lose their colour when canned), and to standardise the appearance of packed foods (for instance the colour of jam differs from one batch to another). Additionally manufactured foods, not made from natural ingredients, are coloured to make them acceptable: a strawberry flavoured jelly requires red colour, and orangeade must obviously be coloured orange.

The Colouring Matter in Food Regulations 1973 permit about 35 groups of colours, shown in the table on page 85. They are permitted in any food, except raw or unprocessed meat, fish, fruit, vegetables, white bread, tea, coffee and milk, which are not legally allowed to be coloured. Brown and wholemeal bread may only be coloured with caramel, and butter and cheese with specified natural colours (like annatto). Nuts may have colour on the shell; methyl violet and burnt umber can be used for marking meat and cheese; and aluminium, silver and gold can be used for cake decorations.

There are three types of colours used in food: inorganic (like carbon black used for example in liquorice); 'natural', or their synthetic equivalent; and synthetic dyes. There are several alternatives for each colour used in different foods; some colours only dissolve in fat, others in water, and others may react with the food. Cakes, salad cream and squash may all contain different yellows.

Coal tar dyes are preferred to natural colours by the food industry—they are less likely to deteriorate and, being highly purified, very small quantities are required—but the harmlessness of some has been questioned. Amaranth for example has recently been banned in America: animal tests implicated it as a cause of cancer and damaging to the foetus in pregnancy. In Britain it is the most commonly used red colour, in red jams, soft drinks, sweets, cakes and ice cream, but has not—as yet—been banned because the American tests are said to have yielded equivocal results.

Changes of colour during cooking and processing of food

Natural plant colours are complex substances, easily altered by heat, light, acids and alkalies. Myoglobin changes from red to brown when meat is cooked: but the colour can be stabilised by **nitrite**. Chlorophyll in green leaves changes from bright to bronze green when cooked. Alkalies in cooking water (like bicarbonate of soda) preserve the green colour but destroy vitamin C. Carotene (in orange and yellow fruits and vegetables) is destroyed at high temperatures in air: apricots turn brown when they are dried unless the **preservative** sulphur dioxide is added. Anthocyanins in red, purple and blue fruits and vegetables turn red in acids (for instance vinegar) and blue in alkaline cooking water. Flavones in white fruits and vegetables (like cauliflower) yellow on

PERMITTED COLOURS IN FOOD

NATURAL COLOURS OR THEIR SYNTHETIC EQUIVALENTS

Curcumin	Found in turmeric
Riboflavin	Vitamin B_2
Cochineal	Extracted from the Coccus cacti beetle
Orchil	Red, extracted from lichens
Chlorophyll	In green leaves
Caramel	Formed when sugar is burnt
Carotenoids	21 are permitted, including *carotene* and annatto, butter colour. They range from yellow, orange, red colours.
Betanin	From beetroot
Anthocyanins	In, for instance, red cabbage. Red in acids, blue in alkalies
Spices	Paprika, turmeric, sandalwood

INORGANIC

Carbon black	Used, for instance, in liquorice
Iron oxide, hydroxide	Red/brown
Titanium dioxide	White

COAL TAR DYES

Solanthrene Blue RS	Blue
Patent Blue V	
Brilliant Blue FCF	
Indigotine	
Black BN	Black
Black 7984	
Green S	Green
Red 2G	Red
Carmoisine	
Amaranth	
Erythrosine	
Ponceau 4R	
Tartrazine	Yellow
Quinoline Yellow	
Sunset Yellow FCF	
Fast Yellow AB	
Yellow 2G	
Orange G	Orange
Orange RN	

cooking in alkaline water: the colour change can be prevented by acids—lemon juice, vitamin C or vinegar.

Another important change which occurs when food is cooked or processed is browning, brought about by three different processes:

Heat—caramel is formed when sugars are heated. It is a permitted colour and added to many foods, including brown bread, marmalade and pickles.

Enzymes—cutting, peeling and chopping of vegetables and fruits releases enzymes from the damaged cells. Flavones in white vegetables and fruits are changed from white to brown in the presence of oxygen in air. Browning of potatoes, bananas, apples, can be prevented by excluding oxygen (by covering with water or adding vitamin C), or blanching (dipping in hot water) or by adding sulphur dioxide.

Maillard reaction—Amino acids—particularly lysine—and sugar combine to form a brown coloured complex when food is heated, and during storage of some processed foods. Although Maillard browning is desirable in cooked foods (for instance the skin of milk pudding, the crust of bread) in others—like milk powder—it is not: the nutritional value of the protein and flavour of the food are adversely affected.

CONVENIENCE FOODS

Manufactured food requiring little additional preparation in the home. For example, meat pies, dehydrated and frozen meals, instant puddings and soups, cake mixes and breakfast cereals.

Convenience foods save time for the consumer, but even allowing for lack of waste, they are generally a more expensive way of obtaining essential nutrients than fresh foods: labour and fuel used in preparation, packaging and transport have to be paid for. Losses of essential nutrients during processing are not much greater than in normal home cooking, but the nutrient content depends on the ingredients used: convenience foods are likely to contain less of the more expensive ingredients (like meat) than a home prepared dish. In Australia, a study demonstrated that convenience foods (like rice with curry or meat sauce) contained only 55% of the protein and 60% of vitamin B_1 on average compared with meals prepared in the home.

An additional reason for taking only moderate quantities of convenience foods is that they contain more food **additives** than fresh or lightly processed foods. As a safety factor many food additives are not permitted in foods eaten in large quantities. Convenience foods are assumed to be eaten infrequently and are thus subject to more lenient additive regulations.

COOKING FATS—lard, dripping, suet, lard substitutes (compound or white fat)

These separated **fats** are concentrated sources of **energy**. All except shredded suet are 99% fat and are virtually deficient in all other essential **nutrients**. Lard, dripping and suet are saturated fats (see fatty acids) and contain **cholesterol**.

Lard is extracted from pig fat, dripping from sheep or oxen bones or fat. Suet is extracted from oxen or sheeps' kidney fat: block suet must contain 99% fat, but shredded

AVERAGE NUTRIENTS IN 25 GRAMS (1 OZ) COOKING FATS

Nutrient	Lard	Dripping	Shredded suet	Lard substitute (white fat)
Energy (kilojoules)	965	965	840	965
(kilocalories)	230	230	200	230
Fat (grams)	25	25	20	25
Cholesterol (milligrams)	18	18	13	a
Vitamin E (α tocopherol milligrams)	0·5	—	—	b
Carbohydrate (grams)	0	0	4	0
Protein, minerals and other vitamins	0 or trace only			

a – none if made from vegetable oils
b – content depends on *oil* used
— – no information available

suet contains up to 15% rice or wheat flours (to keep the pieces apart).

Lard substitutes (compound cooking fats, white fats or shortenings) are made from blends of oils—usually vegetable. They contain no cholesterol but are hardened (or hydrogenated—see Margarine) and therefore also contraindicated in high polyunsaturated, low saturated fat diets.

All cooking fats are permitted to contain **antioxidant**.

COPPER

A trace element, necessary for growth of children and part of many **enzymes** including those needed for the formation of blood and bone.

The table below shows approximate copper contents of foods: liver is the most outstanding source (about 30 grams (1 oz) will supply the adult estimated daily needs). Unrefined cereals (like wholemeal bread) supply more copper than

APPROXIMATE COPPER CONTENTS OF SOME FOODS

Rich sources	Containing more than 1 milligram of copper in 100 grams (3½ oz) food:
	Liver, shellfish, malted milk drinks, cocoa, curry, wheat germ, yeast, fruit gums, brazil nuts
Good sources	Containing 0·4 to 1 milligram of copper in 100 grams of food:
	Wholemeal bread, all bran, puffed wheat, shredded wheat, flaked wheat
	Mushrooms, parsley, broad beans, currants
	Chocolate, liquorice, pastilles, toffee, treacles
	Beef and yeast extract
Moderate sources	Containing 0·1 to 0·4 milligrams of copper in 100 grams food:
	Most other cereals and bread
	Cream, evaporated and condensed milk (undiluted)
	Most meat and fish
	Pulses, most nuts, spinach, broccoli, potatoes
	Red wine
Poor sources	Containing less than 0·1 milligrams of copper in 100 grams food:
	Boiled rice and pasta
	Most cheese, fresh milk, ice cream
	Butter, other fats, oils
	Eggs, pork, eels, salmon
	Most fruits and vegetables
	Sugar (white or brown), soft drinks, boiled sweets
	Spirits, beers, cider, white wine

refined (like white bread): milk, most dairy products and eggs are notably poor sources.

Adults probably require between 1·5 and 2 milligrams of copper per day—an amount which is supplied by a well **balanced diet**, containing a variety of foods. Children require more in proportion to their weight than adults to allow for growth, and more is required during pregnancy to ensure an adequate liver store of copper in the newborn child.

Adults contain between 100 and 150 milligrams of copper, distributed throughout all the tissues of the body but in highest concentration in the liver, brain and kidneys. Copper-containing enzymes are known to be necessary for the release from liver stores of **iron** used in new red blood cell formation (see Anaemia); for the formation of **proteins** in bone, skin and blood vessels; and for the formation of melanin—the pigment in skin and hair. The most vital role of copper containing enzymes is probably the final stages of removal of energy from food and transferring it to 'energy rich' substances (like ATP—see Phosphorus). These are a source of energy in—for example—the synthesis of many complex substances including **phospholipids**.

Adults are unlikely to suffer from a deficiency of copper except when suffering from diseases causing **malabsorption** of nutrients from food. Most foods contain small amounts and a mixed diet (see Balanced diet) is protective. However, infants, who are dependent on one food (milk) for the first few months of life, are more at risk. Normally, copper is transferred from the mother towards the end of pregnancy and held in store in the liver. This store tides the infant over until weaning. Symptoms of copper deficiency—which have occurred in premature babies and those fed for a long time with fresh or evaporated cows milk—include failure to thrive, diarrhoea, anaemia and bone fractures. In some cases the bone fractures resembled those found in battered babies. Replacement of copper in the diet cures the condition. Cows milk contains less copper than breast milk: it is added to some powdered baby milks.

Like other trace elements, copper is toxic in excess. It causes diarrhoea and over the long term accumulates in the body, causing liver damage. Rarely food may be contaminated with copper-containing fungicides. Legal maxima for copper include a general limit of 2 milligrams per 100 grams for most foods: soft drinks may only contain 0.2 milligrams per 100 millilitres.

Two inherited diseases cause copper deficiency and excess in the body. Wilson's disease results in accumulation of copper in the brain and liver. Although previously fatal, this disease can be treated with drugs which remove copper from the body. In Menkes syndrome, too little copper is absorbed from food, causing brain damage (possibly because too little phospholipid is formed) and other symptoms of copper deficiency.

Animal experiments suggest that other trace elements—cadmium, lead (see Poisons in food) and zinc may interfere with the absorption of copper from food. The high ratio of zinc to copper, in cows milk compared with breast milk, may be a contributory factor in copper deficiency in infancy. When given to animals in excess **molybdenum**, another trace element, causes a syndrome resembling copper deficiency. Knock knees and other skeletal abnormalities have been reported in adults and adolescents subsisting on very restricted diets in India where the soil is high in molybdenum and **fluorine**. Low intakes of copper, or high intakes of copper antagonists may be of relevance in **heart disease**: in animals a low copper diet results in a raised blood **cholesterol**.

CREAM

Cream has a similar nutritional value to **butter** in that it contains **fat, vitamins A, D, E**, and is a poor source of protein, calcium and B vitamins. It contains more water and therefore less energy than butter, and small amounts of **carbohydrate**. Compared to margarine and butter it is an expensive source of nutrients: like butter it also contains

cholesterol and a high proportion of saturated **fatty acids**. Summer cream contains more vitamins A, D and E than winter cream.

The nutritional value of creams varies according to their fat content. High fat creams—like clotted and double cream —contain more energy and vitamins A, D and E than single cream. However there is slightly more calcium and protein in single cream. By law, clotted cream must contain 55% of fat; double cream 48%; whipped and whipping cream 35%; sterilised cream 23%; single cream 18%; half cream and sterilised cream (including longlife UHT creams) 12%. Milk contains 3–4% fat. Whipped and sterilised cream may contain **emulsifiers** and stabilisers which must be declared on the label. Whipped cream may also contain sugar; clotted cream, the **preservative** nisin. No other additives are allowed. The table below compares the nutrient content of single and double cream. Sour cream is made from single cream, inoculated with lactic acid producing bacteria, see Yoghurt. It has the same nutritional value as single cream. Imitation cream is made from vegetable oils, emulsifiers and water. It contains no nutrients, apart from fat.

AVERAGE NUTRIENTS IN 50 GRAMS (1¾ OZ) CREAM

Nutrient	Double cream	Single cream
Energy (kilojoules)	970	460
(kilocalories)	230	110
Protein (grams)	¾	1¼
Carbohydrate (grams)	1	1½
Fat (grams)	24	10
Cholesterol (milligrams)	85	40
Calcium (milligrams)	25	40
Vitamin A (microgram equivalents)	210	75
Vitamin D (micrograms)	0·15	0·05
Vitamin E (milligrams α tocopherol)	0·5	—
B vitamins, other minerals	Only small amounts	

— – no information available

CYSTEINE AND CYSTINE

Two non-essential **amino acids**. They are interchangeable in the body: cystine is composed of two molecules of cysteine.

Cysteine and cystine are made in the body from the essential amino acid **methionine** but are also present in food **proteins**. They are required, like other amino acids, for synthesis of new protein needed for growth and repair, and are especially abundant in keratin, the protein in hair. Despite claims for pollen (which contains cystine) taking extra will not arrest hair loss.

Cystine and cysteine in the diet reduces the needs for methionine, and, since almost all the sulphur in the diet is derived from these three amino acids the **sulphur** content is sometimes used as an approximate assessment of the adequacy of a protein. Cysteine is an additive used in new **bread** making processes.

DEXTRINS

Formed when **starch** is subjected to dry heat, for instance in grilling or frying; or to strong acids; or to the action of **enzymes**, for instance in **digestion**, malting of cereals (see Maltose), and bread making. Dextrins are brown, taste slightly sweet, dissolve in water and have the same **energy** value as starch.

Dextrins are formed when bread is toasted, in the crust of bread, in biscuits, bread crumbs and breakfast cereals.

DIABETES—full name Diabetes mellitus

A disease resulting from an insufficient or ineffective supply of the hormone insulin. It is quite common, affecting at least 1 % of the British population. There is a familial tendency to diabetes.

The energy in **carbohydrate** (sugar and starches) reaches the blood stream in the form of **glucose**. Blood glucose (blood 'sugar') rises after a meal and, in response, insulin is secreted in the blood stream from cells of the pancreas—a gland (sweetbreads) situated below the stomach (see also Digestion). Insulin enables body cells to take glucose out of the blood stream and use it either immediately for energy, or to store it for later use in the form of **fat** or **glycogen**. About 4 hours after eating, the blood glucose has returned to normal.

In diabetes, although the glucose passes into the blood stream as before, insulin is not secreted in sufficient amounts, or is made but is ineffective. As a result, the blood glucose remains unacceptably high. The kidneys are unable to conserve the excessive quantities of glucose circulating in the blood stream, and allow it to escape into the urine. Usually, glucose (sugar) in the urine confirms the presence of diabetes.

There are two main types of diabetes, but it may also be caused by other hormone disturbances, some drugs, and liver disease.

Diabetes treated without insulin—or maturity onset diabetes

About 80% of diabetics develop the disease in middle or old age. The majority are obese, and women are more affected than men. See also Obesity and Chromium.

In this type of diabetes, there are often no intense symptoms of thirst or loss of weight (see below), and the disease may only be found as a result of routine urine test. In other cases, the side effects of the disease, such as intense itching around the genital area (caused by the growth of yeasts on the sugary urine), impotence, eye changes, tingling in fingers, loss of feeling in the legs or boils and other infections on the skin may initiate a visit to the doctor. First symptoms may also be experienced as a result of stress, for instance a car accident or operation, but shock cannot cause diabetes—it merely unmasks a latent form of the disease.

Obese people are rather resistant to insulin (the tissues are less responsive to its action) and the pancreas increases production to maintain a normal blood glucose level. When

the pancreas is unable to maintain this high output, overweight people become diabetic. However, often a determined
effort to lose weight, coupled with a low carbohydrate diet,
are all that is needed to restore the blood glucose to normal.
The small amount of carbohydrate allowed should be
divided equally into 3 or 4 small meals a day, taken at regular times. This allows the remaining capacity to produce
insulin to keep the blood glucose within normal limits.
Usually diets containing between 100–150 grams of carbohydrate and 1000 to 1500 kilo**calories** are needed, but treatment depends on individual circumstances and requires
medical and qualified dietetic supervision.

Drugs that lower the blood glucose, either by stimulating
the cells of the pancreas to produce more insulin, or by
enabling body cells to use it more efficiently, are sometimes
needed—particularly for those who are not overweight. A
similar low carbohydrate diet is still required, but insulin
injections are rarely necessary.

Insulin dependent diabetes

Usually this type of diabetes occurs in childhood or early
adulthood, causing serious symptoms which, unless treated,
are rapidly fatal.

Before treatment, there may be a complete lack of effective insulin, and large quantities of glucose spill over into the
urine. Water is drawn out of the body and a thirst develops
to keep pace with the excessive volumes of urine passed.
Additionally, fats are withdrawn from body stores (adipose
tissue), but the large quantities released cannot be used completely for energy purposes and accumulate as acids in the
blood stream, causing a dangerous disturbance of the deli-
cate acid/alkali balance in the body. The loss of incom-
pletely used fat and glucose in urine causes a marked loss of
weight. Other symptoms are pain in the abdomen, vomiting
and weakness. Unless treated, coma results.

Insulin (extracted from animal pancreases) restores the
condition to normal. Unfortunately it cannot be taken by

mouth (it is a protein and would be digested) and has to be injected daily. There are several different types. Some are short acting, effective only for a few hours and necessitating more than one daily injection: others are effective for 24 hours. The diet is adjusted to match the type of insulin needed—more carbohydrate is given when the insulin is at its highest peak of activity. The total daily amount of carbohydrate allowed depends on age and activity: a very active adolescent would require more than an older office worker.

Once established on their regime, diabetics are able to live fairly normal lives, balancing their diet with injections of insulin. However, it is a difficult situation and imbalances can occur. Increases in activity, too little to eat, and too much insulin can all result in a low blood glucose. The brain and nervous tissues are particularly dependent on a supply of glucose and if the blood level falls, their activity is depressed, resulting in drowsiness and eventual coma. A rapidly absorbed form of glucose—usually sugar (sucrose)—is carried by all diabetics treated with insulin and taken when the first symptoms occur. These precautions are not necessary for diabetics treated by diet alone.

After the discovery of insulin in 1922, a 'free' diet with daily adjustments of insulin was the treatment of choice for all diabetics. Later it was realised that late onset diabetics are often resistant to the action of insulin, and that the regime could result in overweight and wide fluctuations of the blood glucose level. For these reasons, free diets have been largely abandoned in favour of treatment with diet alone, or diet with insulin or drugs. Overweight and blood sugar are easier to control if a strict dietary regime is followed, and there is probably less risk of complications (eye changes, nervous degeneration, kidney disease, **atherosclerosis** affecting the limbs and heart) in both types.

Once weight has been stabilised at or below normal, it is possible to vary the diabetic diet by using an exchange system for carbohydrate. Provided the total amount of carbohydrate at each meal is kept constant, and as medically

prescribed, foods containing an equivalent amount of carbohydrate may be eaten in place of those on the diet sheet.
Most clinics use a system of 10 grams of carbohydrate
exchanges—for instance if 50 grams of carbohydrate are
allowed at one meal, 5 portions (5 × 10 grams) of food may
be chosen. Some foods containing 10 grams of carbohydrate
are shown under **carbohydrate**. A comprehensive list is
available from the British Diabetic Association, 3 Alfred
Place, London WC1. In view of their increased risk of **heart
disease**, diabetics are probably well advised to avoid large
helpings of **fat**, especially saturated fats like butter, cream,
and fat on meat.

A chance of finding glucose in the urine is not always indi-
cative of diabetes. In young people, and sometimes during
pregnancy, the kidneys may fail to conserve glucose completely, even though the blood glucose is normal. Other
people may absorb glucose very quickly from food (particularly if part of the stomach has been surgically removed) and
it is lost in the urine whilst the blood level is temporarily
high. It is normal for these reasons to perform a confirma-
tory test—the Glucose Tolerance Test—which measures the
body's response to a test dose of carbohydrate (usually a
glucose drink). Blood is withdrawn every half hour for 2
hours after taking the dose and its content of glucose measured. Diabetics have a characteristic response. Prolonged
fasting and very low carbohydrate **slimming diets** (for example the 'quick weight loss diet' and Atkins diet) can also
cause a diabetic curve; normal food should therefore be
taken for several days before undergoing a glucose tolerance
test.

DIET

Daily food intake (or an average of food eaten in a week or
longer). Specifically applied to the quantity of **nutrients** in
food, related to bodily needs. Dietetics, practised by dieti-
tians, is concerned with the provision of a combination of

foods which will supply an optimum intake of nutrients for individuals or groups of people. See Balanced diets, Therapeutic diets.

DIGESTION

The dismantling of **protein, fats** and **carbohydrate** in food into particles small enough to be transferred into the blood stream.

One of the smallest animals, amoeba, forms a sac of cell tissue around its food and secretes digestive juices into the sac. Complex digestive systems have evolved in man and higher animals, but the basic process is the same. **Enzymes** are used to dismantle proteins to **amino acids**, carbohydrates to **glucose** and other simple sugars, and fats to **fatty acids** and glycerol. To a minute extent, similar changes can be brought about by heat and acids used in cooking.

The digestive system consists of a series of tubes (beginning at the mouth and ending at the anus) whose contents are kept moving from one section to the next by sequential contractions of circular muscles surrounding the tubes. Digestive juices, containing enzymes, manufactured by glands whose outlets are open on to the digestive system, are mixed with the food as it is forced along. The whole system is controlled by hormones and the nervous system.

Digestion begins as soon as food is put into the mouth. The primary role of the mouth in digestion is the mastication of food, but saliva, secreted by the salivary glands, also contains an enzyme which begins the digestion of starch to **maltose** and **dextrins**. Bread for instance, if chewed for a long time, begins to taste sweet.

After swallowing, food is pushed down into the stomach, where hydrochloric acid and an enzyme, pepsin, are secreted by glands lining the stomach wall. Muscular contractions mix the food and secretions to a porridge consistency, and pepsin begins to dismantle proteins. Small quantities of the mixture are forced out of the stomach at intervals, and six hours after eating a meal it has all passed into the first section of the small intestine, the duodenum. A valve normally prevents food from passing back into the stomach.

Bile, made by the liver and held in the gall bladder, and pancreatic juice, from the pancreas, pass down the bile duct to the duodenum as food enters. The salts in bile emulsify (finely divide) fat into microscopic globules and pancreatic enzymes initiate the digestion of fats, continuing the digestion of proteins and carbohydrates.

The next stage is the entry into the rest of the small intestine—the ileum and jejunum. This tube is about 27 feet long, and its surface area is made even greater by millions of finger-like villi which project into the centre of the tube. The villi themselves have microvilli, microscopic hair-like borders, projecting from them. Here the nearly digested food is absorbed by the surface cells of the villi, and enzymes in the cells complete the breakdown of fats, proteins and carbohydrate.

The main products of digestion are glucose, fructose, galactose, amino acids and an emulsion of microscopic fat particles (chylomicrons). All except the chylomicrons are passed directly to the blood stream and are transported to the liver, whereas most fats travel, as chylomicrons, in lymph vessels which run parallel to the intestine and are later released into the blood stream. Glucose and fructose already present in food (for example in fruits) do not have to be digested and pass directly into the villi. Alcohol and most minerals, and vitamins are also small enough to be absorbed intact. Some minerals however form insoluble salts which cannot be absorbed—see Phytic acid and Oxalic acid. Fat soluble vitamins A, D, E and K, are transported with fat and therefore cannot be absorbed in the absence of bile.

Usually at least 95% of carbohydrate, fat and protein is absorbed out of food in the small intestine, and eight to nine hours after eating a meal, the fluid residue has passed into the large intestine. Here most of the water is absorbed. Undigestible substances (fibre, insoluble salts), undigested nutrients, residues of bile and digestive juices, mucus, dead cells from the lining of the digestive system, and bacteria are later evacuated from the body as faeces. Sixty to 90% of the weight of the stools is water, and up to half the dry weight is bacteria that have multiplied in the large bowel on the fibre and small undigested food residues.

The digestive system is complex and subject to many disorders. Some require hospital treatment and/or modification of diet. See also Dyspepsia, Peptic **ulcers**, Malabsorption, Coeliac disease and Allergy.

DISACCHARIDES

Two combined simple sugars (monosaccharides). The most important are sucrose (table sugar), lactose (milk sugar) and maltose (malt)—see individual entries.

During **digestion**, disaccharides are split to their component monosaccharides which are absorbed into the blood stream. Sucrose is divided to **glucose** and **fructose**, lactose to glucose and **galactose**, and maltose to glucose. Dilute acids can also divide disaccharides—for example **invert sugar**, in honey and used in the food industry, is formed from sucrose in this way.

DYSPEPSIA

Nausea, heartburn, discomfort or distension felt after taking meals. It is usually described as severe 'indigestion', but in the majority of cases, **nutrients** in food are completely digested in the normal way—see Digestion.

Dyspepsia can be a symptom of any disease of the digestive system but it is often brought about by worry, overwork, hurried meals, and over-indulgence in cigarettes and alcohol. It is usually alleviated by simple measures, but medical advice should be sought, especially if these fail and when it occurs in previously unaffected middle-aged people.

Advice for sufferers from dyspepsia

(1) All food must be chewed slowly and thoroughly. Water or other beverages should not be taken with food because it encourages swallowing before food has been properly chewed.

(2) Beverages should be taken between or after meals, and strong coffee, tea and alcohol should not be taken on an empty stomach.

(3) Cigarettes should be avoided.

(4) Large fatty meals should not be taken. Small, frequent meals should be taken at regular times during the day. It is advisable to rest if possible before and after meals.

(5) A bland diet (see below) may help in acute attacks for a few days, but apart from avoiding fried food, pickles and highly spiced meals the diet should be normal over the long term. For most cases of dyspepsia the advice above and avoidance of worry and overwork, together with adequate sleep, are more effective than elaborate dietary regimes. Some authorities recommend that the diet should be low in **carbohydrate**, particularly avoiding sugar (**sucrose**): others recommend a high **fibre** diet.

Bland diet—a typical menu

Small helpings of all food should be taken.

Breakfast—Cornflakes, Rice Krispies, with little sugar and milk

Poached, scrambled or boiled egg; or white fish or grated cheese; or lean, tender ham

Toast, with little butter, and jelly jam or marmalade (no pips or peel)

Weak tea or coffee

Lunch (or evening meal)—Lean, tender meat (no sausages, fatty or twice cooked meat); or white fish baked in milk

Mashed, jacket or boiled potatoes

>   If wanted, boiled cauliflower tips or marrow
>   Milk pudding

Tea   —Plain biscuit (Marie type), weak tea

Supper (or lunch)—Cream soup
>   Grated cheese; boiled, poached or scrambled
>   egg; or lean, tender ham
>   Toast; or mashed, jacket or boiled potatoes
>   If wanted, tomatoes with skin and pips removed
>   or tender lettuce leaves
>   If wanted, junket or jelly or mousse

Before bed—Warm milk or plain biscuits

Rose hip or blackcurrant syrup or fresh or tinned fruit juice
(strained and diluted with water) should be taken if the diet
is continued for more than 2 to 3 days.

EDIBLE GUMS

Extracts from seeds and exudates (saps) of trees, used as
thickeners and stabilisers (see Emulsifiers) in the food
industry and for confectionery. Examples are carob gum,
gum guar (from seeds) and gum arabic, gum tragacanth and
gum ghatti (exudates). They are not split to their component
parts by digestive enzymes and contribute to dietary fibre.
  Seaweed products (alginates, agar, Carragheen, which see)
and dextrin must be specified as such on food labels although
they are also gums. Dextrin used to be called British gum.

EGGS

Eggs are a nutritious food. Their protein is of high quality,
containing a well balanced proportion of essential amino
acids and the only nutrient they are completely lacking is
vitamin C. Eggs are virtually carbohydrate free.
  A standard egg weighs about 60 grams (2 oz). One third of
its weight is the yolk—a concentrated source of nutrients,

containing 30% **fat** (including **lecithin** and **cholesterol**), 16% **protein**, iron, B **vitamins** and vitamins A, D and E, and other **minerals**.

Just over half the weight of an egg is white—mostly water (88%), and protein (9%). The white is virtually fat-free and contains smaller amounts of minerals and B vitamins than the yolk. Vitamin B_1 (thiamin) and vitamins A, D, and E are lacking. Two proteins in the white—conalbumin and avidin —interfere with the absorption of **iron** and **biotin** into the blood stream. More iron is absorbed if eggs are eaten with vitamin C (for example in fruit juice) but otherwise, despite their fairly high content of iron, eggs are not a particularly good source. Cooking inactivates avidin—all the biotin in cooked eggs can be absorbed, but none from raw.

The remainder is shell and shell membranes. The shell is mainly composed of calcium salts, but the white and yolk are not good sources of calcium.

Except for vitamins A and D (see below) the nutritional composition of eggs is remarkably constant irrespective of the hens' diet or breed. There is no significant nutritional difference between white and brown eggs—in general white hens lay white and brown hens brown eggs. Other eggs have approximately the same composition as hens'—duck eggs have a marginally higher fat content.

The average vitamin A content of an egg is 85 micrograms, but a dark coloured yolk does not imply that it has a high carotene (pro vitamin A) content: other pigments in grass and hens' feed (which do not have vitamin A activity) also colour the yolk. Vitamin D cannot be measured precisely, but free range hens have access to sunlight and their eggs are likely to contain more, particularly in summer, than others. Usually vitamin D is added to the diet of battery and deep litter hens: 100 grams (2 small) of these eggs contain about 2·5 micrograms of vitamin D—the adult daily **recommended intake**.

About 14,000 million eggs are produced each year in the UK. In 1971, 85% were battery, 10% deep litter and less than 6% free range. A recent government investigation into the nutritional composition of different types of eggs found that free range and deep litter contained nearly twice as much vitamin B_{12} and slightly more folic acid on average than battery eggs. Other nutrients measured were practically the same.

Eggs are eaten in relatively small amounts and do not contribute greatly to the nutrient content of the average diet, despite being a good source of most. It is a pity, in view of their cheapness and ease of preparation, that they are often regarded only as a breakfast food. In 1973, only about 4 eggs (8 oz) were eaten per person per week, compared with 37 oz per person per week of meat and meat products. Their highest contribution to the average intake of nutrients was 16% of vitamin D.

Eggs keep relatively well because the shell, membranes and white are designed to protect the develping chick and its food store (the yolk) from infection by bacteria and other micro-organisms. Should micro-organisms penetrate the layer of mucin on the shell (which hardens when the egg is laid), the shell, or the membranes, their growth is retarded by the alkalinity of the white. Additionally iron and biotin (necessary for the growth of bacteria) are inactivated by conalbumin and avidin. Other proteins prevent the growth of some viruses and are able to liquefy bacteria.

When an egg is laid, the contents cool to air temperature and shrink, drawing in air through the porous shell. The egg and shell membranes part, forming the air space at the blunt end. Washing or extensive handling remove the mucin coat, exposing shell pores and hastening the loss of water from the inside. If eggs do have to be washed it should be done in warm rather than cold water, otherwise more shrinkage and drawing in of micro-organisms occurs. It is best to store eggs in covered containers which retard water evaporation and flavour penetration through the shell.

During storage of eggs, several changes occur. A fresh egg has very little thin white, and the round yolk is supported on

Average nutrients in a standard (grade 3 or 4) egg weighing 60 grams

Nutrient		Notes
Energy (kilojoules)	410	—
(kilocalories)	100	—
Protein (grams)	7	—
Fat (grams)	7	All in yolk
Cholesterol (milligrams)	300	All in yolk
Iron (milligrams)	1·5	Mostly in yolk
Calcium (milligrams)	35	Mostly in yolk
Phosphorus (milligrams)	130	Mostly in yolk
Potassium (milligrams)	80	—
Sodium (milligrams)	80	Mostly in white
Magnesium (milligrams)	7	—
Other minerals	Small amounts	—
Vitamin A (equivalents)	85	All in yolk
Vitamin D (micrograms)	1	a All in yolk
Vitamin E (milligrams α tocopherol)	1	—
B vitamins:		
thiamin (milligrams)	0·06	All in yolk
riboflavin (milligrams)	0·3	—
nicotinic acid (equivalents)	1·8	—
pyridoxine (milligrams)	0·15	Mostly in yolk
biotin (micrograms)	15	Mostly in yolk
pantothenic acid (milligrams)	1	Mostly in yolk
folic acid (micrograms)	15	b Mostly in yolk
vitamin B_{12}	1	b

a – varies depending on amount of sunshine hen has received, or amount of vitamin D added to feed

b – greater in deep litter and free range eggs

a thick circular wadge of white. As the egg becomes older, water moves from the white to the yolk and a stale egg has a flat yolk surrounded by a thin runny white. The white and yolk become more alkaline as the gas carbon dioxide diffuses out through the shell, and the white proteins and muscular supporting strands (chalazae) are thinned. The yolk moves away from its central position, but if eggs are stored blunt end uppermost the yolk sustains less damage because it is cushioned against the air space. The air space also increases in size and very stale eggs float in water. Eventually the yolk breaks and the proteins are broken down, causing the bad egg smell of hydrogen sulphide.

These changes are retarded by refrigeration, but most eggs sold in shops are class A (old name first class); which are not legally permitted to be refrigerated or preserved by other methods. They should be clean (but not washed) with a minimum of blood spots. The average age of eggs sold is about a fortnight—boxes stamped 'extra' must not be more than 7 days old. Class B eggs may be preserved (see below) or refrigerated. Class C eggs are for food manufacturers only.

Eggs may be preserved by dipping in oil (see Mineral hydrocarbons) to close the shell pores or by storing in carbon dioxide which preserves the acidity of the white. Eggs stored under these conditions eventually lose some vitamin B_1 and vitamin B_{12}, but there is little loss of other nutrients on storage until the eggs turn bad. Liquid egg and dried egg is also prepared but is only available to food manufacturers and caterers.

On cooking, the egg proteins are coagulated (see Protein) and are made firm and easier to eat. Lightly boiled eggs leave the stomach more quickly than raw or hard boiled eggs but all egg is equally well digested. The almost complete absence of residue left after digestion, may be the basis for the belief that eggs are 'binding'. Provided that the egg is not overcooked, the protein suffers no damage. Vitamins A, D and nicotinic acid are completely retained but up to 16% of vitamins B_1 and B_2, depending on the length of time the egg is cooked, may be lost.

A green ring around the yolk in a hard boiled egg is due to a reaction between sulphide (released from the white proteins) and iron in the yolk. It can be prevented by boiling for the shortest possible time and quick cooling in cold water. This reduces the amount of sulphide formed and draws it out of the egg away from the yolk. Stale eggs are more likely to develop green rings than fresh because the protein has already partially decomposed.

Eggs are used extensively in cooking, to thicken and bind foods. Overcooking causes the white to shrink and squeeze water out (for example in an overcooked baked egg custard or scrambled egg). Eggs are also useful **emulsifiers**, used in mayonnaise and meringues.

Egg products and substitutes

Custard powder is a substitute for custard made with egg. It is cornflour (see Maize) with added colouring. Custard made with powder is inferior in its nutrient content to egg custard.
Golden baking powder is baking powder with added yellow colour (see Raising agents). The carbon dioxide produced raises cakes, simulating the air retaining properties of eggs. It has no nutritional value.
Salad cream and mayonnaise are made from oil, egg, starch, gums, flavours and colours. Mayonnaise is not legally required to contain more egg or oil than salad cream.

ELECTROLYTES—see also Water

Certain substances dissolved in the body fluids.

The body is composed of millions of cells (sacs of protoplasm enclosed in membranes) bathed in fluid. This extracellular fluid resembles sea water in having as its chief electrolytes sodium and chlorine (common salt). The extracellular fluid is made up of one part blood plasma and four parts of interstitial fluid. However, cells also contain water— the intracellular fluid, which occupies twice the volume of extracellular fluid. The chief electrolytes in intracellular fluid are potassium, magnesium, phosphates and proteins: it contains very little sodium and chlorine.

Most of the **energy** expended by the resting body is used in maintaining the differential between the two fluids: **enzymes** cannot function efficiently if the electrolyte composition varies outside normal limits. Maintenance at a constant level of the total concentration of electrolytes is the job of the kidneys in particular. The balance can be upset if electrolytes are not absorbed from food (for example in diarrhoea) and in certain diseases, particularly of the kidneys.

EMULSIFIERS AND STABILISERS INCLUDING THICKENERS—see also Additives

Emulsifiers and stabilisers are used extensively in the food industry in foods like bread, cakes, biscuits, meringues, cooking fats, sweets, soft drinks, jams, sauces, and instant puddings. Emulsifiers and the incorporation of air into a liquid—as in ice cream; fat into water—as in synthetic cream and salad dressings; and water into fat—as in margarine. Stabilisers prevent the mixture from separating, but some emulsifiers are also stabilisers.

Substances like oil and vinegar (mainly water) can be made to mix by beating with a whisk or shaking in a bottle, when the oil is divided into globules and disperses evenly throughout the vinegar. However, the globules of oil soon coalesce and rise to the top of the vinegar if the dressing is allowed to stand. But if the oil is beaten into an emulsifier, such as egg yolk, it is divided into much smaller globules. The emulsifiers and stabilisers in the egg yolk surround the globules, preventing them from joining together. At the same time, because of the chemical structure of the emulsifiers, the globules of oil are repelled by each other and attracted to the water in the vinegar. The resulting emulsion (mayonnaise) is thick and does not separate.

Emulsifiers and stabilisers are used in most processed foods and the manufacture of new convenience foods has been facilitated by the development of new emulsifiers. Instant puddings and soups, whips, coffee whiteners and powdered drinks contain emulsifiers to prevent caking and allow the powder to dissolve completely and quickly without lumpiness. They are also used to thicken puddings and other foods without the necessity for heat. Emulsifiers added to

aerated soft drinks prevent particles of fruit clumping together and forming sediment, and trap air in the foam at the
top of the drink. In bread and rolls they retard staling, in
cakes and cake mixes they can substitute for egg, and they
are added to processed cheese and peanut butter to prevent
separation of the fat. In ice cream and mousses, stabilisers
hold globules of fat around pockets of air, allowing them to
keep their foamy structure after thawing—in contrast to
home made ice cream. In chocolate, emulsifiers prevent
stiffening during manufacture and bloom on storage—caused
by sugar (**sucrose**) crystallising out of the fat, sugar and
cocoa emulsion. In cooking fats, emulsifiers prevent droplets
of water coalescing and spluttering out of the fat when it is
heated for frying food. Emulsifier is sometimes added to rice
to prevent starch from leaching out of the grain and causing
stickiness. Instant potato also contains emulsifier.

There are many emulsifiers, each with a different chemical
structure which determines the combination of foods it will
emulsify. The use of the majority is controlled by the Emulsifiers and Stabilisers in Food Regulations 1975, which permit 57 specified substances. The permitted list includes 21
miscellaneous substances, like **edible gums, lecithin, agar** and
alginates; 30 chemically modified fats (see triglyceride) and
7 chemically modified forms of **cellulose**. There are no restrictions on their use, except that milk, flour and cream (other
than whipped or sterilised) may contain none. Certain specified foods, like bread and soft cheese, may only contain
specified emulsifiers. The amount added depends on the
physical properties of the food—some cake mixes and baking fats contain considerable amounts, but generally small
quantities are required (between 0·5 and 4 grams per 100
grams ($3\frac{1}{2}$ oz) of food). **Starches** can also be chemically
modified to make a wide variety of emulsifiers but—at
present—their use is not controlled by law. Substances like
egg yolk, **dextrin**, proteins, glucose syrup and malt extract
which are also used as emulsifiers are also not controlled
because they are classified as foods.

ENERGY

Many people think of energy as good, and calories as bad (fattening): but it is important to remember that they are effectively one and the same thing. **Calories**, strictly kilocalories, are units for the measurement of energy. Both kilocalories and the more general energy units—kilo**joules**—are used in this section.

Like all living things, the human body is in a state of constant change. Cells are continually being broken down and require energy to repair themselves. About half the daily intake of energy is used for this and other processes of 'self maintenance'. The rest is used for muscular movement (activity) during the day, and for growth.

Energy is released from food to meet daily needs. Over the short term a diet too high or low in energy is harmless. The discrepancy is balanced by adjustment of body stores: if food contains more energy than required it is stored—mostly as fat (see Lipid)—and weight is gained. If food supplies too little, fat is withdrawn and weight is lost. Over the long term, major departures from normal weight must be corrected. Diets low in energy are required for **obesity** and diets high in energy for **starvation**.

There is normally sufficient energy in store (as fat and **glycogen**) to cope with immediate needs. It should be stressed that eating extra energy does not stimulate the body into greater activity, or ensure an 'energetic' or zestful life—though many advertisements (particularly for sugar, sweets and glucose) give the opposite impression. Over the long term, eating more than daily needs can reduce zest: the excess weight gained is tiring. Most 'energetic' people are of normal weight: they eat sufficient to supply their needs, but are not burdened with fat.

Energy in food

All foods contain energy (none are slimming) but their energy content varies considerably depending on the form in which the energy is held, and the amount of water in food. Approximate energy contents are shown below: cooking fats and oils are highest in energy, vegetables contain the least. See individual entries for more information.

APPROXIMATE ENERGY IN 25 GRAMS (1 OZ) PORTIONS OF SOME FOODS

		kilojoules	kilocalories
Energy-rich foods	Cooking fats, oils, butter, margarine, cream cheese	770–990	185–235
	Biscuits, cakes, dry cereals, nuts, sugar, cheese, fatty meat, double cream, chocolate, dried milk	350–670	85–160
Moderate sources	Lean meat, fatty fish, eggs, bread, single cream	150–350	35–85
Low in energy	White fish, most fresh fruit and vegetables, milk	Below 140	Below 35

In 1973 the average British diet (excluding sweets, alcohol and meals bought outside the home) contained 2400 kilocalories (10MJ) per day. Cereals supplied the most energy (30%), followed by meat, eggs and fish (nearly 20%), fats (15%), milk and cheese (14%), sugar and preserves (10%), and fruits and vegetables (10%).

The energy in food is held in the form of fat, **carbohydrate** (sugar and starch), **protein** and **alcohol**. Fats contain the most energy: each gram (1/28th of an ounce) contains approximately 9 kilocalories (38kJ). Foods which are mostly fat supply more energy than those with little—thus fatty meat is higher in energy than lean, biscuits higher than bread. Carbohydrates and proteins each supply about 4 kilocalories (17kJ) per gram, and alcohol about 7 kilocalories (29kJ) per gram.

Most foods contain **water**: foods with a high percentage of water supply less than those which are more concentrated. Thus weight for weight bread contains less energy than flour, fruit less than sugar. Plants also contain **fibre** which is unaffected by digestive **enzymes** and probably supplies little energy to the system.

It is difficult, when leading a sedentary life, to reduce the amount of energy in food sufficiently to avoid overweight, and at the same time ensure an adequate intake of all other

essential **nutrients**. For this reason, 'empty calorie' foods—
sugar, alcohol, some fats—should be avoided.

Energy needs

Individual energy requirements vary widely: some people
use up—on average—three times more energy than others.
Most of the differences are due to variations in activity—at
work and in leisure. Other factors are body weight, age and,
to a lesser extent, climate. The four factors are related in a
complex way and **recommended intakes** for groups of people
are based on an average (reference) man and woman. Few
individuals conform to the average, but without a lengthy
series of tests it is not possible to predict exactly how much
energy an individual requires. For adults, a constant body
weight (with no gains or losses of fat) over the long term
remains the only practical method of ensuring that energy
needs are being met with the right amount of energy in food.

Energy needs of the average man and woman—the average
man and woman weigh 65 kilograms (10 stones) and 55 kilograms (8½ stones) respectively, are aged between 20 and 39
years and are healthy. Each day they spend 8 hours at work
—in a moderately active job or housework—and 9 hours
asleep. Of the remaining 8 hours, 6 are spent in recreation—
sitting or taking light exercise—and two are spent more
actively—in for instance, cooking, washing, cleaning or
sport. Each day they use up 3000 kilocalories (12·5MJ) and
2200 kilocalories (9·1MJ) respectively:—

| Period | | Energy used | |
		megajoules (MJ)	kilocalories
Asleep	Man	2	500
	Woman	1·75	420
At work	Man	5	1200
	Woman	3·6	880
Rest of day	Man	5·5	1300
	Woman	3·75	900
Total	Man	12·5	3000
	Woman	9·1	2200

Effect of activity

Activity has the greatest influence on energy used during the day and recommended intakes for energy are based on occupation. Moderately active workers include most factory workers, housewives, postmen, bus conductors. Sedentary workers—for instance shop workers, bus drivers, office workers, most professional workers—weighing 64 kilograms could use up 300 kilocalories (1·25MJ) less than the average man. Very active workers—for instance miners and steel workers—could use up 600 kilocalories (2·5MJ) more at work than the average man.

During sleep—apart from the work of the lungs, heart and digestive system—no muscles are being used. The body is at rest. The average man uses about 1 kilocalorie (4·2kJ) per minute—or 60 kilocalories (250kJ) per hour—the energy contained in one small slice of bread.

Immediately on rising, the energy used increases (muscular work is required to maintain posture) to about 1·75 kilocalories (7·3kJ) per minute for the average man when he is standing still. More movement increases the amount used—see table overleaf. The recommended intake includes an estimated average energy expenditure of 1300 kilocalories (5·46MJ) whilst not at work or asleep.

Effect of body weight

The energy used during sleep is approximately equal to that required for self-maintenance—the Basal Metabolic Rate (see Metabolism) and depends on the number of active cells in the body. Fat cells (which are food stores) are relatively inactive, and consequently an obese person of the same weight (65 kilograms) as the average man will use up slightly less energy whilst asleep. Women—with more fat cells—also use less energy during sleep than men.

During activity, use of energy is proportional to body weight: it takes more energy to move a heavier weight against the forces of gravity. For instance, the average man uses

3·7 kilocalories (15·5kJ) per minute when walking at 3 mph, but a 75 kilogram (12 stone) man uses 4·0 kilocalories (16·7kJ) per minute. Despite the greater energy cost of movement, an obese person may use less energy during the day than one of normal weight: the extra burden of fat is tiring, and reduces activity.

Effect of age

To allow for growth, infants, children and adolescents require more energy in proportion to their weight than

ENERGY USED UP IN VARIOUS ACTIVITIES BY AN AVERAGE 65 KG MAN AND 55 KG WOMAN (1)

| | Approximate average energy used per minute | | | |
| | by man | | and woman | |
	kilo-joules	kilo-calories	kilo-joules	kilo-calories
Asleep in bed	4·5	1·1	3·8	0·9
Sitting quietly (for instance watching TV)	5·8	1·4	4·8	1·2
Standing quietly	7·3	1·8	5·7	1·4
Cooking	8·8	2·1	7·1	1·7
Light cleaning	13·0	3·1	10·5	2·5
Moderate cleaning (for instance polishing)	18·0	4·3	14·6	3·5
Walking at 3 mph	15·5	3·7	12·6	3·0
Sedentary work—office	7·5	1·8	6·7	1·6
—lorry driving	6·7	1·6	—	—
Light industry—garage repairs, laundry work	17·2	4·1	13·4	3·2
—electrical industry	15·1	3·6	7·9	1·9
Recreations:				
Seated—like knitting	10·5	2·5	8·3	2·0
Light—like golf, sailing	10·5–21	2·5–5	8·3–16·5	2·0–4
Moderate—like dancing, tennis, riding	21–31·5	5–7·5	16·5–25	4–6
Heavy—like athletics, football	31·5+	7·5+	25+	6+

(1) Taken from WHO Tech. Rep. Ser. no 522 1973

adults. Thus a child of 5 needs nearly twice as much energy (weight for weight) than an adult. See Appendix for recommended intakes. Extra energy is required during pregnancy and breast feeding to fulfil the needs of the growing baby. An extra 200 kilocalories (840kJ) (in for example a $\frac{1}{2}$ pint of milk) is recommended daily in the last 6 months of pregnancy and an extra 500 kilocalories (2·1MJ) (in $1\frac{1}{4}$ pints of milk) daily when breast feeding. The elderly generally use less energy than young adults: activity and the number of active cells in the body are reduced with increasing age.

Effect of climate

Humans maintain their body temperature at or around 37°C—best suited to the activity of enzymes—with heat continually dissipated during **metabolism**. Normally, this heat is continually lost by radiation from the skin to the surrounding cooler air. In hot climates, less heat needs to be generated to maintain the body temperature and, to keep cool, most people further reduce their activity. The average man might have to reduce his food intake by 150 kilocalories (620kJ) (contained in for instance two large slices of bread) or more if he moved from a temperate to a hot climate. In contrast, a very cold climate does not always increase the needs for energy to maintain the body temperature: the amount of heat lost from the skin is reduced by increased clothing and home heating.

ENZYMES

Proteins that act as biological catalysts, accelerating the speed of chemical reactions in living tissues. Reactions that would normally require great heat, strong acids or many years can take place almost immediately with the aid of enzymes. They are essential for the existence of life.

Enzymes are formed inside each cell, according to genetic instructions inherent in cells and specific to each species. Some enzymes work outside cells (for instance those required for **digestion** of food) but most are held inside.

Only minute amounts of enzymes are needed to bring

about reactions—in some cases an enzyme can transform a million times its own weight of one compound (substrate) into another. They need a strictly defined acidity or alkalinity in their surroundings and work best at normal body temperature.

Enzymes are specific (each enzyme reacts with one specific substrate or a group of substrates closely resembling each other) and used to be named after the substrate they caused to react. Thus the digestive enzyme that splits lactose (milk sugar) to form **glucose** and **fructose** is called lactase. In modern classification, enzymes are numbered according to the type of reaction they control and are also given a chemical name if their structure is known. Most of the millions of enzymes thought to exist in nature remain to be purified, named and classified.

The activity of an enzyme can be inhibited—thus slowing or stopping a reaction. Many drugs work on this principle, although often the enzyme involved is not known. See also Inborn Errors of Metabolism.

Some enzymes are unable to function unless another, non protein substance is associated with them. These are called coEnzymes and about 20 are known—many contain **B vitamins.** In addition, enzymes may need small amounts of **minerals**—cofactors—to function. Molybdenum, iron, copper, manganese and zinc are known cofactors.

Enzymes are present in foods or added to foods as microorganisms (yeasts, bacteria and fungi). They are responsible for many of the changes in texture, flavour and colour when foods are stored and prepared. Some of the changes are undesirable (for instance the browning (see Colour) of potatoes, and the destruction of vitamin C) and the enzymes responsible are destroyed during processing of food. In other cases, enzymes take part in the preserving of food, and in making it palatable. For instance, when meat is hung, enzymes partially degrade the protein fibres, so tenderising it. Enzymes in yeast and flour are necessary for bread making; enzymes in bacteria are necessary to convert lactose to lactic acid in yoghurt making; and enzymes in fungi are necessary for the flavour of some cheeses.

Many enzymes can now be extracted from micro-organisms (as yet, they cannot be synthesised) and a wide range is used in the food industry. Starch splitting enzymes are used to make glucose syrup from starch, and sugar (sucrose) splitting enzymes are used in soft centre chocolates (see Invert sugar). **Pectin** splitting enzymes can be used to make vegetable purées. As yet, the use of enzymes is not controlled by law, but a Government Committee (see Additives) is intending to review their use in food in the near future.

EXTRACTS—from meat and yeast

Most meat extract is a by-product of the corned beef industry. Successive batches of meat are heated in a tank of

AVERAGE NUTRIENTS IN 10 GRAMS (⅓ OZ) OF EXTRACTS

Nutrient	Marmite	Bovril	Oxo (red)
Energy (kilojoules)	2	35	85
(kilocalories)	1	8	20
Protein (grams)	Trace	2	½
Fat (grams)	Trace	Trace	½
Carbohydrate (grams)	0	0	2½
Iron (milligrams)	0·7	1·2	2·5
Potassium (milligrams)	280	160	45
Sodium (milligrams)	440	450	2650
Magnesium (milligrams)	20	17	Trace
Phosphorus (milligrams)	190	130	35
Other minerals	Small amounts		
Vitamins A, C	0	0	0
Vitamin E	—	—	—
B vitamins:			
thiamin (milligrams)	0·3	—	—
riboflavin (milligrams)	0·6	0·4	—
nicotinic acid (milligram equivalents)	6	5	—
pyridoxine (milligrams)	0·4	—	—
pantothenic acid (milligrams)	0·6	—	—
folic acid (micrograms)	130	—	—
biotin (micrograms)	10	—	—
vitamin B_{12} (micrograms)	0·1	—	—

— – no information available

water when flavours and B vitamins are leached out of the meat. Fat is skimmed off and the extract is concentrated or dried by boiling. Brand products, including Bovril and Oxo, have added flavours (like **hydrolysed protein** and yeast extract), salt, starch and caramel. With the exception of **thiamin** (which is largely destroyed by the heat) meat extracts contain **vitamin B, minerals** (including iron), and some protein. They are rich in **salt**.

Yeast extract is made by autolysis (a process of self-digestion). **Enzymes** produced by dead cells destroy cell walls, liberating the contents. In brand products, salt, spices and vegetable extract are added to the concentrated yeast extract. Yeast extract is a good source of B complex vitamins but poor in protein. Its very high salt content makes it unsuitable for babies.

The table opposite shows the nutrient content of 10 grams of Marmite, Bovril, and Oxo. One cup contains about 10 grams, and about 3 grams are used on a slice of bread.

FAT—see also Lipid, Butter, Margarine, Cooking fat, Oils, Cream

Fat is the highest **energy** yielding **nutrient**. Foods which are nearly all fat—like vegetable oils—provide more energy weight for weight than any other food.

Although generally used to describe concentrated fatty foods like fat on meat and separated fat like butter, the term 'fat' applies equally to the non-visible fat contained in nearly all foods. Lean roast beef and eggs, for instance, contain about 12% of their weight as fat: peanuts about half. The exceptions—fat-free foods—are fruits, most vegetables, sugar, egg white, and the majority of beverages. Foods containing very little fat—low fat foods—include white flour and bread, skimmed milk, cottage cheese, very lean poultry, veal and rabbit, and some shell fish and white fish.

The table on p. 118 shows portions of foods containing approximately 10 grams of fat, but there are variations in the non-visible fat content of some fresh foods at different times of the year and between different species. Herrings are leaner after the spawning season and Jersey milk has a higher fat content than Friesian. **Meat**, cooked foods and **cheese** also have widely differing fat contents—cream crackers contain over twice as much fat as water biscuits for instance, and Edam cheese about half the fat content of Stilton. The fat content of some foods is also altered during processing— wholegrain cereals contain a small amount of fat, but most of it is removed during milling.

PORTIONS OF SOME FOODS CONTAINING AN AVERAGE OF
10 GRAMS FAT

	Food	Weight containing 10 grams fat
Separated fats	Lard, oil, dripping	10 grams ($\frac{1}{3}$ oz)
	Butter, margarine	12 grams ($\frac{1}{2}$ oz)
	Double cream	20 grams ($\frac{2}{3}$ oz)
Invisible fats	Peanuts, fried streaky bacon	20 grams ($\frac{2}{3}$ oz)
	Chocolate, cheddar cheese	30 grams (1 oz)
	Lean roast loin of pork	50 grams (1$\frac{2}{3}$ oz)
	Plain biscuits, baked herring	75 grams (2$\frac{2}{3}$ oz)
	Lean roast beef, eggs	80 grams (2$\frac{2}{3}$ oz)
	Ice cream	90 grams (3 oz)
	Oatmeal (raw)	115 grams (4 oz)
	Raw liver, roast chicken	130 grams (4$\frac{2}{3}$ oz)
	Milk	270 grams ($\frac{1}{2}$ pint)

The prime function of fat is to provide energy in the diet. Apart from the essential nutrient **linoleic acid**, the liver is able to make all the fat the body requires from **carbohydrates** and **protein**, provided these are eaten in sufficient quantities. Weight for weight, fat supplies over twice as much energy to the body as carbohydrates (sugar and starches): each gram of fat supplies about 9 kilo**calories** and each gram of carbo- hydrate or protein about 4 kilocalories.

The amount of fat used in a diet depends on food pre- ferences and customs but in general it is a highly prized constituent of meals. Most people in Western countries

would find a diet that supplied less than 50 grams a day unpalatable. The majority of flavours and colours are dissolved
in fat; it lubricates food making it easier to swallow; and it
prolongs the length of time food is left in the stomach—
increasing the satiety value of meals (very fatty meals may
cause indigestion because they remain in the stomach for a
long time).

In Britain the average person eats about 120 grams of fat
a day, which provides nearly half the energy in the diet.
About half the average intake is obtained from meat, fatty
fish, eggs, milk, cream and cheese. A further third is eaten as
butter, margarine and other separated fats; and pastries,
cakes and biscuits contribute about 10 grams. The rest is
supplied by small amounts of fat in cereals.

The incidence of ischaemic **heart disease** in a community
can be correlated with the average individual fat content in
the diet. In general industrialised countries (one exception is
Japan) eat more fat and have a higher rate of death from
heart disease. Less affluent countries eat less fat on average
and have a lower death rate from heart disease. There are
other risk factors, but it is beneficial to take only moderate
fat in the diet especially for those with **gout** or leading
sedentary lives with a tendency for overweight. This can
most easily be done by eating less fried food, cream, butter,
margarine, fat on meat, and rich creamy sauces and sweets.
See also Lipid and Cholesterol.

Very low fat diets are inadvisable, except when medically
prescribed, because there is a danger of depleting the body
of fat soluble vitamins A and D. However, low fat diets
(containing about 40 grams a day) may be needed as part of
the treatment of some diseases of the liver and digestive
system—see Malabsorption—and occasionally prior to the
removal of gall stones. For modified fat diets see Lipid.

In low fat diets separated fats are severely restricted and
fatty fish and meat, ordinary milk, most cheese, cream,
pastry, cakes and biscuits made with fat, chocolate and
confectionery, and ice cream are usually forbidden. Fat free

foods, especially sugar and starch, should be taken liberally: low fat diets are invariably low in energy and can result in severe weight loss. Fruit juice with extra glucose is often readily accepted. Special purified oils—medium chain triglycerides (MCT) (see Fatty acids)—can be used for frying and to make margarine (used for spreading and baking). MCT oil is expensive (though available in some circumstances on prescription) but a valuable source of extra energy in a low fat diet.

FATTY ACIDS

Structurally, a fatty acid is a chain of carbon atoms, each with hydrogen atoms attached. At the end of the chain is an acidic group of atoms, able to combine with **glycerol**. Three fatty acids attached to one molecule of glycerol is called a simple fat (chemical name **triglyceride**). Nearly all **fat** in the diet is composed of triglycerides—see also Lipid.

Short chain fatty acids contain 2 to 6 carbon atoms, medium chain 8 to 10 and long chain 12 to 30. Most fatty acids are long chain, but some short chains—like butyric acid—are found in milk fat, butter and cream. Medium chain fatty acids are extracted from oils to make medium chain triglycerides (MCT)—water miscible fats used as a source of extra energy in some low fat diets.

Fatty acids containing their full quota of hydrogen atoms are called *saturated fatty acids*. Animal fats and hardened (see Margarine) fats tend to contain more saturated fat than plant and fish and be solid at room temperature. Three saturated fatty acids—lauric, myristic and palmitic acids—have been found to raise the blood **cholesterol** level of humans, but other saturated fatty acids have little effect. The three cholesterol raising fatty acids are the most commonly occurring saturated fatty acids. Palmitic acid is very common: in animal fats and palm oil it makes up to 35% of all fatty acids, and up to 17% in other plant oils and fish oils.

Fatty acids with less than their full quota of hydrogen atoms are called *unsaturated fatty acids*. The hydrogen is replaced by a double bond between the carbon atoms. Most plant and fish oils contain a greater proportion of unsaturated fatty acids than animal fats and are liquid at room temperature. Fatty acids with only 2 hydrogen atoms missing (and therefore only one double bond) are called *monounsaturated fatty acids*. Monounsaturated fatty acids have no effect on the blood cholesterol. In nature, oleic acid—a monounsaturated fatty acid with 18 carbon atoms—is the most common fatty acid, most fats and oils contain 30 to 65% of their total fatty acids as oleic.

Fatty acids with four or more hydrogen atoms missing (and therefore two or more double bonds) are called *polyunsaturated fatty acids* (or PUFA or polyunsaturates). Polyunsaturates have been found to lower the blood cholesterol, but they are less effective in lowering the blood cholesterol than are saturated fats in raising it. Linoleic acid (18 carbon atoms and 2 double bonds) is one of the most abundant fatty acids in plant oils, but it also occurs in small

APPROXIMATE PROPORTIONS OF FATTY ACIDS IN SOME FATS (AS A PERCENTAGE OF TOTAL FATTY ACIDS)

Fat	Saturated[a]	Stearic[b]	Mono-unsaturated[c]	Poly-unsaturated[d]
Butter, milk, cheese, cream	40	10	30	4
Beef	30	20	45	2
Lamb	30	15	50	5
Pork and bacon	28	12	50	8
Chicken	25	5	45	20
Fish (herring)	20	Trace	25	50
Wheat germ	20	1	10	65
Corn oil	10	3	35	50
Avocado	20	1	50	15
Olive oil	15	2	70	10
Cocoa butter	25	35	35	2
See also oils, margarine and nuts.				

a – Total lauric, myristic and palmitic acids
b – Stearic acid is saturated fatty acid but has less effect on blood cholesterol than lauric, myristic and palmitic acids
c – Mainly oleic acid – the most common fatty acid
d – Mainly linoleic acid, but fish fats contain other polyunsaturated fatty acids

amounts in animal fats. Fish oils tend to contain large
amounts of other polyunsaturated fatty acids.

Oils and fats with a high ratio of saturated to polyunsaturated fatty acids are avoided in cholesterol lowering
diets, and (provided there is no need to reduce weight) oils
with a low ratio are taken instead, see Cholesterol and Oils.

Essential fatty acids are polyunsaturated fatty acids with
an arrangement of double bonds that cannot be imitated by
the body. **Linoleic acid** is the initiator of this family and must
be taken in the diet. All other fatty acids can be synthesised
by the body and are not essential **nutrients**.

FIBRE

Dietary fibre is the part of leaves, stems, roots, seeds and
fruits of plants that is not digested by **enzymes** produced in
the human digestive system (see Digestion) and therefore
increases the bulk of the daily bowel motion. It does not
include tough fibres of meat or fish—these are usually fully
digested. Other names for dietary fibre are roughage, bulk,
and unavailable carbohydrate.

Nearly all vegetable foods contain fibre. Of the **cereals**,
foods made from the wholegrain, containing the **bran**, supply
more than milled. Thus wholemeal **bread** and wholewheat
breakfast cereals (like puffed and flaked wheat) contain
more than white and brown breads. The fibre is not completely removed in milling, and white flour (in cakes, biscuits
and white bread) may contribute significant amounts to the
diet. In **vegetables** and **fruits** usually the tough parts and the
skin, pips and peel contain more than the fleshy inner parts.
Pounding and chewing do not reduce the fibre content of
food, but sieving and treatment with enzymes (to make
juices and purées) do. Other sources are some food additives,
like **edible gums, agar** and **alginates. Cocoa** and **nuts** are good
sources. Meat, fish, fats, oils, eggs, cheese, milk, and sugar
contain no dietary fibre.

The estimated dietary fibre contents of some foods are

shown in the table facing. Most dietary fibre is in the form of complex **polysaccharides**. Unlike **starch** (also a polysaccharide) they cannot be broken down to their component parts by digestive enzymes in humans, but may be fermented by bacteria which normally inhabit the large bowel. Another constituent of dietary fibre is lignin, the main substance in wood. The proportions of the different constituents of dietary fibre vary between foods, particularly between cereals and vegetables. This variation may be as important as the total dietary fibre content of foods.

Dietary fibre is not a **nutrient** and, until recently, it was regarded as an unimportant part of the diet, apart from helping to prevent constipation. In some circumstances it was said to be harmful because of its supposed irritant action on the digestive system and, in some foods, high **phytic acid** content. High fibre diets also reduce the amount of protein and **energy** absorbed out of food. However, it has been claimed that several disorders of the digestive system, including diverticulitis, and cancer of the large bowel, appendicitis, and gall stones; varicose veins, haemorrhoids, ischaemic **heart disease**, tooth decay, **obesity**, and **diabetes** are all caused by lack of cereal fibre in the diet. The possibility that so many common diseases are caused by a deficiency of fibre has stimulated a great deal of research, but until more is known about the way the many complex constituents of fibre behave in the body, most of the claims will remain unproven.

Theories relating a deficiency of fibre to these diseases are based on evidence obtained partly by studying the incidence of disease and type of diet, and partly by measuring the time food takes to pass through the digestive system.

All the cited diseases are more common in most industrialised countries and in other communities eating a 'refined' (low in fibre content) diet. They are rare in rural communities in African countries where a traditional high fibre diet is still eaten. It is claimed that their incidence increased in Britain and other wheat eating countries after the introduc-

DIETARY FIBRE IN 100 GRAMS (3½ OZ) PORTIONS OF FOODS

Foods	Unavailable carbohydrate,[a] grams
Almonds, fresh coconut	10–15
Brazil nuts, peanuts, chestnuts, cobnuts, walnuts	5–10
Blackcurrants, dates, redcurrants, blackberries, stewed prunes, raspberries	7–8·5
Bananas	3·4
Gooseberries, fresh figs, strawberries, eating apples,[b] apricots,[b] pears,[b] plums, oranges, stewed rhubarb	2–2·5
Tangerines, stewed cooking apples, cherries	1·5–2
Peaches,[b] pineapple, cantaloupe melon	1–1·5
Grapes,[b] yellow melon, grapefruit[b]	0·5–1
Orange juice, strained[b]	0
Spinach,[b] peas, baked beans	5–6·5
Brussels sprouts, broad beans, broccoli	4–5
Leeks, mustard and cress, spring greens	3·5–4
Watercress, runner beans, carrots, spring onion	3–3·5
Swede, cabbage, cauliflower, boiled lentils, mushrooms, parsnips, beetroot	2·5–3
Celery, lettuce,[b] onions, tomatoes, asparagus tips	1·5–2
Potatoes, radishes	1–1·5
Marrow, cucumber[b]	0·5–1

	Dietary fibre,[c] grams	Crude fibre,[d] grams
Wholemeal flour	11	2
Brown flour	—	0·6
White flour	2·5	Less than 0·2
Rye flour	17·3	1·5
Bran	—	10

a – probably equivalent to dietary fibre

b – skin/and/or pips, pith, outer leaves and stems removed. If these are eaten, fibre contents will be higher

c – the most accurate estimate of dietary fibre, but few figures are available (the method has only recently been developed). These are taken from Southgate (1969) *J. Sc. Fd. Agric.* **20** 331 and Frazer (1958) *J. Sc. Fd. Agric.* **9** 136

d – the 'official' method for measuring fibre, used by analysts and in *bread* regulations. It is much quicker than method c, but underestimates most of the different components

tion of roller mills in the 1880s when white **flour** and bread became cheap and widely available. Coupled with a decline in the amount of bread eaten, the cereal fibre in the majority of diets in these countries is said to have been greatly reduced. In Britain, during the Second World War, the incidence of some of these diseases decreased, and this decrease is attributed to the higher content of fibre in wartime bread. At this time, the extraction rate (see Flour) of flour was increased to save shipping space.

During **digestion**, nutrients are absorbed out of food in the small bowel. The residue—containing dietary fibre—then travels along the large bowel until it reaches the rectum. By giving a 'marker' with food, the time taken between food entering the mouth and stools leaving the anus can be measured and is called the transit time. Transit times have been found to be shorter in communities eating a high fibre diet than in those eating a low fibre diet, probably because fibre absorbs water and swells, increasing the bulk of the intestinal contents, which may move faster.

Too little fibre is one cause of constipation (see below) but even if bowel habits are 'regular', low fibre diets are claimed to be harmful because residues are comparatively small and hard, and remain in the large bowel for a long time. In an effort to propel the small hard residues along, the large bowel muscles are said to increase in size to exert more pressure, when the increased pressures set up cause herniations in the large bowel wall (diverticulae)—rather like a blown tyre—which may become infected (diverticulitis). It is also claimed that residues can be forced into the appendix (a small pouch off the large bowel) which may result in appendicitis, and that the stomach can be forced up through the diaphragm, causing hiatus hernia—a painful condition where acid is regurgitated into the oesophagus (food pipe). The increased pressure is also said to be transferred to the veins which collect blood from the legs and anal area, causing engorgement of the veins with blood and eventual haemorrhoids and varicose veins. Some bile

salts are degraded by bacteria in the large bowel and—
although none have so far been detected—carcinogens
might be formed. A low fibre diet is claimed to increase the
risk of bowel cancer because, with slower moving residues,
more carcinogens would have time to be formed, and they
would be in contact with the bowel surface for longer.

It is generally agreed that low fibre diets are the usual
cause of constipation, though irregular habits and lack of
exercise are also important. An excessive **water** intake does
not prevent constipation, though an early morning cup of
tea (or other hot drink) may aid evacuation by stimulating
contraction of the bowel muscles. Continued use of purgatives and laxative salts is harmful: eventually the bowel
muscles may lose their tone, resulting in constipation or
diarrhoea in old age. The best way of increasing the fibre
content of the diet is to replace white bread with wholemeal
(not brown, wheatmeal or granary, which are legally allowed
to contain much less fibre). Plenty of vegetables and fruit
should also be eaten—salads are no better than cooked
vegetables. Alternatively, bran or cellulose or other proprietary bulk remedies can be taken. Six teaspoons (20
grams) daily of bran is usually sufficient for most people and
it can be eaten with breakfast cereal, soup, fruit salad,
yoghurt or on bread with jam.

Most hospitals now treat diverticulitis with a daily
supplement of 20 to 30 grams of bran (taken as 2 to 3
teaspoons three times a day) and a high fibre diet as outlined
above. The majority of people are relieved of their symptoms
of abdominal pain, flatulence and sometimes mucus and
diarrhoea (these are also symptoms of other bowel disorders
and should always be medically investigated before treatment with bran). However, although it has been demonstrated that a low fibre diet can cause the condition in
animals, it is not entirely certain that a low fibre diet is the
cause in humans. It is also uncertain whether low fibre diets
cause appendicitis and other related disorders. At present a
high **fat** diet is thought to be more important in predisposing

to bowel cancer. This type of diet requires more bile for its digestion and may result in the formation of more carcinogens.

The relevance of fibre to the other cited diseases—obesity, diabetes, tooth decay, heart disease and gall stones—is even more uncertain. A high fibre diet reduces the amount of energy absorbed out of food in the small bowel, probably because there is less time for digestion and absorption of nutrients into the blood stream. It is claimed that the small deficit in the amount of energy absorbed—over the years— might be sufficient to avoid obesity and its attendant complications (like diabetes developing in middle age). Low fibre diets are also claimed to have a low satiety value: it is suggested that it is easier to overeat when the diet is refined. A high fibre diet needs more chewing. These apparent benefits may be outweighed in practice—compared to white bread, wholemeal may be eaten with more butter, jam and other spreads to make it palatable. In its unrefined form (for instance sugar cane and fruit) it is also suggested that fibre causes less dental caries: fibre is said to have a cleansing action on the teeth. It would be more beneficial to cut down on jam, sugar and sweet consumption than increase the fibre content of the diet.

Ischaemic heart disease and gall stones may be related to the ability of fibre to bind bile salts, which are made in the liver from **cholesterol**. Fibre, by carrying them away from the body in the stools, is claimed to reduce the body content of cholesterol. A high blood cholesterol is one risk factor in heart disease, and gall stones contain excess quantities. So far, high cereal fibre diets have reduced the blood cholesterol in animals, but not in human experiments. See Heart disease.

Although most of the benefits claimed for high fibre diets remain controversial, increasing the fibre content of the diet by replacing fats and sugar with wholegrain cereals, potatoes, fruit and vegetables is probably beneficial. The diet is then more likely to contain a higher ratio of essential nutrients to energy. However, if much bran is taken the **phytic acid**

content of the diet will be increased and it is important to
ensure a good intake of **calcium** (of which milk, cheese and
yoghurt are rich sources) and **vitamin D**.

FISH

A valuable source of good quality **protein** and **minerals**. All
fish contain **phosphorus**, sea fish are important for their
iodine content, and fish with small bones that cannot be
separated from the flesh are a source of **calcium**. All fish
contain B **vitamins**, fatty fish are a source of fat soluble
vitamins A, D and E, and most fish are **carbohydrate** free.
White fish and some fresh water fish like bream, perch and
pike, contain minimal fat and a high proportion of water.
With their low **energy** and high protein content they are ideal
foods for slimmers. Compared with meat, white fish are as
good a source of B vitamins, but a poorer source of iron.
The slightly fattier white fish (like halibut) sometimes contain
small amounts of vitamin A, but usually the flesh of white
fish contains no fat soluble vitamins. The livers of white fish
are however rich sources of vitamins A, D and E. Roes,
including caviar, are rich in protein and B vitamins—
particularly **thiamin**—and contain about as much **iron**,
weight for weight, as lamb or pork. About one third to a
half of thiamin is lost when roes are cooked.
Fatty fish and some fresh water fish like salmon, eel and
trout are between 5 and 30% fat, depending on season and
species. Their fat is high in polyunsaturated **fatty acids**.
They are generally fattest in late summer and early autumn,
and leanest after spawning in early spring. In nutritive value
fatty fish are superior to white, containing more energy, fat
soluble vitamins A, D and E and iron. Herrings contain as
much iron weight for weight as lamb, sardines as much as
beef. Their bones, which are often eaten, are rich sources of
calcium. In view of their cheapness, eels and mackerel are
particularly valuable foods. Eels and brislings are rich
sources of vitamins A and D; some reportedly contain at

least 25 micrograms of vitamin D in a 100 grams ($3\frac{1}{2}$ oz) portion—enough to fulfil the adult **recommended intake** for 10 days.

Shell fish, including crustaceans (like crabs, shrimps) and molluscs (like mussels, whelks) contain twice as much salt as other fish, traces of carbohydrate (**glycogen**), and about the same protein, fat and energy as white fish. Only the flesh of crustacea is eaten and their vitamin and other mineral contents are similar to white fish. Molluscs are eaten whole (including their organs) and are better sources of iron and vitamin A. Without shell, cockles, whelks and oysters contain as much **iron**, weight for weight, as beef; winkles as much as liver. Some oysters contain vitamin C.

Despite their nutritional virtues, only 4·7 ounces of fish were eaten in the average British diet (excluding meals taken outside the home) in 1973, compared with 36·5 ounces of meat. Their greatest contribution to the average intake of nutrients in the diet was about 19% of vitamin D—mostly from fatty fish. One reason for the unpopularity of fish may be the risk of staleness. Over half the samples bought in a *Which?* survey in May 1972 from fishmongers were stale. Fish bought on Friday was usually staler than that bought on Monday, but fish bought from seaside towns was usually fresh.

Staling is retarded by keeping the catch on ice and by adding **antibiotics**, but fish are generally a few days old when bought and should be eaten as soon as possible. Bought fish should have bright eyes, red gills and plenty of firmly attached scales. They should smell fresh (slightly seaweedy) and have firm creamy (not yellow) white flesh. Their markings, for instance spots on plaice, should be clear. Shell fish should be heavy in proportion to their size.

Unlike meat, fish contains no connective tissue and consequently is never tough—unless overcooked when the flesh becomes rubbery and dry. Only sufficient heat to coagulate the protein, turning the flesh white, is required. On boiling, nearly half the minerals—including iodine of

AVERAGE NUTRIENTS IN 150 GRAMS (5 OZ) PORTIONS OF SOME FISH (VALUES FOR STEAMED OR BAKED FISH UNLESS OTHERWISE STATED)

Nutrient	White fish				Herring	Fatty fish			Shellfish		Roe
	Cod	Haddock	Plaice	Whiting		Mackerel	Canned salmon	Canned sardines	Mussels	Prawns	Cod
Energy (kilojoules)	515	610	580	565	1190	1180[a]	865	1850	550[b]	655[b]	805
(kilocalories)	125	145	140	135	285	280[a]	205	440	130[b]	155[b]	190
Protein (grams)	27	33	27	30	26	30[a]	30	30	25[b]	32[b]	36
Fat (grams)	1¼	1¼	3	1½	20	15[a]	10	35	3[b]	2¾[b]	5
Cholesterol (milligrams)	75	400	—		125	140[a]	55	210	75[c]	225[c]	450[c]
Iron (milligrams)	0·75	1·05	0·9	1·5	2·4	1·8[a]	1·95	6	20	1·65[b]	3·45
Calcium (milligrams)	20	80	55	65	85	45[a]	100	615	295[b]	215[b]	20
Potassium (milligrams)	540	485	420	450	350	625[a]	480	650	140[b]	390[b]	200
Phosphorus (milligrams)	365	350	370	285	490	420[a]	425	1025	495[b]	525[b]	600
Magnesium (milligrams)	30	40	35	45	35	50[a]	45	60	40[b]	65[b]	12
Sodium (milligrams)	150	180	180	190	95	230[a]	805	1180	315[b]	2385[b]	110
Iodine (micrograms)	Sea fish, average 120; fresh water, average 5; salmon, average 50										
Other minerals	Small amounts										
Vitamin A (microgram equivalents)	0	0	0	0	70[e]	70[e]	140	45	90	0	—
Vitamin D (micrograms)	0	0	0	0	35[e]	25[e]	20	10	0	0	3[e]
Vitamin B:											
thiamin (milligrams)	0·09[e]	0·09[e]	0·09[e]	0·09[e]	0·05[e]	0·15[e]	0·05	0	0·08[e]	0·08[e]	2·25[e]
riboflavin (milligrams)	0·15[e]	0·15[e]	0·15[e]	0·15[e]	0·45[e]	0·75[e]	0·15	0·3	0·23[e]	0·15[e]	1·5[e]
nicotinic acid (milligram equivalents)	9[e]	9[e]	9[e]	9[e]	9·5[e]	12[e]	15	12	5[e]	10[e]	8
pyridoxine (milligrams)	0·3[e]	0·3[e]	0·3[e]	0·3[e]	0·7[e]	1·0[e]	0·45	0·24	0·05[c]	0·15[c]	—

AVERAGE NUTRIENTS IN 150 GRAMS (5 OZ) PORTIONS OF SOME FISH

Nutrient	White fish				Fatty fish				Shellfish		Roe
	Cod	Haddock	Plaice	Whiting	Herring	Mackerel	Canned salmon	Canned sardines	Mussels	Prawns	Cod
B vitamins:											
pantothenic acid (milligrams)	0·3e	0·3e	0·3e	0·3e	1·5e	—	0·75	0·75	0·75e	0·45e	—
biotin (micrograms)	15e	15e	15e	15e	—	—	15	7·5	15e	—	—
folic acid (micrograms)	9	—	—	—	—	—	—	15	—	—	—
vitamin B_{12} (micrograms)	1·5e	1·5e	1·5e	1·5e	15e	—	3	—	—	—	—
Vitamin E (α tocopherol milligrams)	0·8e	0·8e	0·8e	0·8e	—	—	—	15	0	0	—
Vitamin C (milligrams)	0	0	0	0	0	0	0	0	0	0	45e

a – fried; b – boiled; c – values for oysters, shrimps and caviar respectively; d – salt added; e – value for raw fish; — – no information available which fish is one of the few reliable sources—leach into cooking water. All can be recovered if the liquor is used for sauce. Baking, grilling and frying cause little loss of nutrients.

Fish can be frozen, canned or cured. Freezing causes no loss in nutritive value, though some B vitamins and flavours are lost in the drip on thawing. To avoid these losses, small pieces of fish should be cooked without thawing beforehand. Up to 75% of thiamin is lost when fish is canned (compared with 30 to 50% losses in home cooking). Canned sardines contain none. The other nutrients investigated are stable to canning. Kippers contain no thiamin, but in general cured fish contain more nutrients, weight for weight, than fresh: salt, used in curing, removes water. Smoking can be cold, when the fish remains raw (kippers), or hot, when the fish is cooked (smoked trout).

Of the many fish products, only the composition of fish cakes and spreads are controlled by specific regulations. Legal minimum weights of fish in these products are shown opposite. Fish fingers (not subject to control) usually contain 50 to 70% of fish. Fish products may contain most **additives**: **colours** (brown in kippers, yellow in smoked fillet); **flavours**; **emulsifiers**; and miscellaneous additives, like sequestrants added to prevent struvite formation in tins of shell fish. Struvite crystals are harmless but may be mistaken for glass. Other additives, not yet subject to permitted lists, like smoke solutions, may also be used. Smoke solutions and **polyphosphates** avoid the loss of water which usually occurs with traditional curing processes. **Antioxidants** and **preservatives** may be present in ingredients, and although not directly added, will be 'carried over' to the finished food. Frozen fried fish for instance will contain antioxidant.

Fish is rarely a cause of food poisoning—fresh fish becomes unpalatable when subjected to conditions favouring the multiplication of food poisoning bacteria. It is also usually cooked immediately before eating (unlike some meat foods). Hot smoked fish, which is not cooked before eating, is a possible danger if it is not kept in a refrigerator—

particularly smoked trout which, in contrast to sea fish, can be contaminated with the botulism toxin—see Poisons in food.

Most episodes of poisoning by shell fish are due to **allergy** but 'wild' molluscs can be contaminated with sewage and should be gathered with care. They also accumulate a potent nerve toxin formed by plankton which sometimes multiplies in summer time in the sea (colouring it red), causing poisoning of wild life and humans. Although outbreaks of mussel poisoning are quite common elsewhere, the first recorded case in England occurred in 1968. About 80 people were affected. It is probably wise to avoid gathering shell fish in summer months.

Fish is the main source of methyl mercury, the most toxic form of mercury (see Poisons in food). The average mercury content of fish landed in Britain is quite low and about 1 lb a day would have to be eaten to exceed the adult provisional tolerable intake. Tuna and shell fish contain more than other fish. However, fish caught in some British estuaries (particularly the Thames and Mersey) has a much higher mercury content—1 lb a week would exceed the tolerable limit.

COMPOSITION OF FISH PRODUCTS

Food	Legal minimum fish content
Fish cakes	35%
Fish pastes and spreads	70%
Fish pastes and spreads labelled 'requires grilling'	None
Potted fish	95%
Potted fish and butter	96% (fish and butter)
Fish paste with one other main ingredient	80%

FLAVOUR

A mixture of many chemicals in minute quantities, recognised by taste and aroma. It is important in stimulating appetite and identifying foods, but the way in which flavour is perceived is only partly known. **Zinc** and perhaps **vitamin A** are necessary for flavour perception.

The taste buds on the surface of the tongue are sensitive

to sweet, sour (acid), bitter and salt. However, aroma (which stimulates the sense of smell) is much more important in distinguishing foods and promoting appetite. An apple tastes the same as an onion when the nose and sinuses are blocked. Heat intensifies aroma—the chemicals are volatile —but overcooking can drive most of the flavour out of food.

In cooked foods, flavour depends partly on that of the raw ingredients (which may be affected by breed, maturity, and environment); changes brought about by cooking; and substances rich in flavouring chemicals—flavouring agents— added to impart their own flavour. About 1500 flavouring agents are used by the food industry. They may enhance the flavour of the food, replace losses in processing, or flavour manufactured foods (for instance jellies, instant puddings, margarine, soft drinks, **novel proteins**). There are four main groups of flavouring agents—foods (about 30); **herbs** and **spices** (about 150); essential oils, extractives and distillates (about 250); and synthetics (about 1000 specified chemicals). **Salt**, **sweeteners** and **monosodium glutamate** are also used.

Examples of foods with flavouring properties are **hydrolysed protein**, yeast **extract**, anchovies, vegetables and soy sauce. The essential oils, extractives and distillates are essences—concentrations of the flavouring chemicals in fruits, herbs, spices and vegetables obtained by for instance pressing, distillation, or treatment with a solvent. Ginger oil, oil of lemon, gum benzoin, vanilla extract are examples.

Natural flavouring agents may not be suitable for some processed foods and are often difficult to obtain. For instance, a quarter of a million crocus stamens have to be gathered for each 1 lb of saffron, and vanilla orchids must be artificially pollinated the day the flower opens and the bean later carefully picked, fermented and matured. Very little essence can be extracted from the original material (often less than 1%) and the extraction processes are very complex. These difficulties have prompted the development of synthetic flavours.

Mixtures of synthetic chemicals can stimulate a natural

flavour. Examples of synthetic flavours that can be made are apple, apricot, banana, blackberry, brandy, rum, butter, cherry, chocolate, almond, raspberry, strawberry, cranberry, date, garlic, grape, honey, lemon, peach, pineapple, vanilla, even cheese (Cheddar or Roquefort) and bacon. However, synthetic flavours cannot exactly mimic a natural flavour and are usually inferior. Although some synthetics are identical to their natural counterparts, comparatively few of the thousands of flavour components that exist naturally are known and can be synthesised. Examples of synthetics also occurring in a natural flavour are acetaldehyde (apple), benzaldehyde (almond), diacetyl (butter), diallyl disulphide (garlic), vanillin (vanilla), citral (lemon), and probably undecalactone (peach). Synthetics are particularly useful as fortifiers—when added to a natural flavouring agent they increase its potency.

Flavours are not currently subject to specific government regulations, apart from the need to declare their presence in most prepacked foods (see Labelling) and the general provisions of the Food and Drugs Act 1955 which makes it an offence to sell food injurious to health. In 1965 the Food Standards Committee recommended that there should be a list of 17 prohibited flavours. These are rarely used and probably unsafe—for instance saffrole (in Sassafras) and coumarin (in Tonka bean) cause liver damage in rats fed very large doses. A permitted list (of substances thought to be safe) was not feasible because of lack of information. Although the prohibited list would have been in force only until a permitted list could be introduced, it was not accepted by the Ministry of Agriculture, Fisheries and Food. The Food Standards Committee was asked to prepare a permitted list but the necessary toxicological tests (see Additives), which are the responsibility of the food manufacturers wishing to use the additives, have not yet been completed.

FLOUR

Usually prepared from **wheat**, though other **cereals**, pulses and nuts may also be ground into flour. Wheat flour has a high **gluten** content and can be made into leavened **bread**. Cornflour (see Maize) is virtually pure **starch**.

Flour can be milled to different extraction rates. In 100% extraction flour (wholemeal), all the grain is used and 100 lb of flour is obtained from 100 lb of wheat. The flour is dark because it contains the outer layers of the grain—**bran** and germ, rich in **minerals, fibre** and **vitamins**. In 70% extraction flour, 70 lb of white flour is obtained from 100 lb of grain. All the outer layers are removed, leaving only the inner part (endosperm). Brown flours can be between 85 and 95% extraction (containing most of the germ and some or most of the bran) or they may be a blend of white flour with sufficient bran added to comply with legal requirements, see Bread.

All flour used to be made by grinding wheat between stones. Some of the bran could be separated off by forcing (bolting) through cloth but the 'white' flour produced was probably equivalent to an 80% extraction flour. It was cream coloured due to the presence of most of the germ and some of the bran. It was very expensive: most people ate wheaten (made with coarser cloths and containing more bran) or household (wholemeal, or wholemeal with rye) breads. In the 1880s roller mills were introduced. These separated the bran and germ, enabling millers to produce cheap white flour of 70% extraction or lower. This white flour was inferior nutritionally, but more popular and replaced brown and wholemeal in the average diet. The health of working people (who were still dependent on bread for most of their essential **nutrients**) probably declined as a result. Since 1953, three 'token' nutrients (**iron**, and two B vitamins—**thiamin** and **nicotinic acid**) must legally be partially replaced (up to the level of 80% extraction flour) in white flour. Although white flour remains nutritionally

inferior to wholemeal, the significance of the difference in a
normal **balanced diet** remains uncertain. See Bread and
Fibre. These nutrients are also usually added to all brown
flours—though this is not strictly necessary if the flour is of
a high extraction rate (about 80%). No added nutrients are
permitted in wholemeal flour. **Calcium** (see Phytic acid) is
also added by law to all flours except wholemeal.

No **additives** are permitted in wholemeal flour, but eleven
bleaches and improvers are allowed in all other flours. These
hasten changes which occur naturally when flour is stored—
the colour of white flour changes from cream to white and
most strengthen gluten, enabling a taller, lighter loaf to be
made. Two—chlorine and sulphur dioxide—are only
permitted in cake and biscuit flours respectively. See table
overleaf.

Although wholemeal bread is not at present allowed to
contain improvers, in 1974 the Food Standards Committee
recommended that **ascorbic acid** and **cysteine** (numbers 4
and 6) be permitted. The Food Standards Committee also
recommended that all bleaches and improvers should be
subject to regulations specifying the maximum amounts
allowed in flour. Safety tests for food additives are supposed
to be carried out by the industry with an interest in their
continued use in food. In 1960 the Committee specified that
long term toxicity tests should be made on all bleaches and
improvers but was disturbed to find in 1974 that these had
not been completely carried out for five (numbers 1, 3, 7, 8
and 9 in the table). So far as is known these are harmless in
the amounts normally eaten but, unless tests are completed
by the next review and there are no adverse effects, these
additives are likely to be banned.

Technological advances (including the use of additives
and modern mills) enable millers to produce standardised
blends of flours for specific purposes, despite natural variations in gluten (protein) content and baking properties. In
general, flours intended for bread, pasta and puff pastry
contain more gluten and are said to be 'stronger' than cake

Additives Permitted in All Flour Except Wholemeal

Number	Name	Action	Notes
1	Chlorine dioxide	Improver, strengthens gluten. Bleach	Used for about 80% of flour. Destroys vitamin E
2	Benzoyl dioxide	Bleach	Used for 60–70% of flour, usually with 1, 3, 7 or 8. Destroys vitamin E. Converted to benzoic acid on baking—see *preservatives*
3	Potassium bromate	Improver, strengthens gluten	One of most widely used improvers, usually together with 1, 2, or 4. Leaves a residue of bromide but too little to affect wakefulness
4	Ascorbic acid	Improver, strengthens gluten	Chemical name for vitamin C. Used together with 3 and 6
5	Azodicarbonamide	Improver, strengthens gluten	Recently permitted. Useful for new bread-making processes, together with 3. Converted to harmless substance in bread (biurea)
6	L cysteine hydrochloride	Improver, weakens gluten	An *amino acid*. Recently permitted. Useful in new bread-making methods, together with ascorbic acid (4)
7 and 8	Ammonium persulphate, Potassium persulphate	Improvers, strengthen gluten	Very small amounts used, mainly in Scotland and N. England
9	Chlorine	Improver, strengthens gluten. Bleach	Only allowed in cake flours. Used in less than 1% of total flour, and not in retail flour. Used for cakes with a lot of sugar (high ratio) for instance Angel Cake because it softens texture
10	Sulphur dioxide	Improver, weakens gluten	Destroys vitamin B_1 and only permitted in biscuits. Used in about 15% of biscuit flour—chiefly hard sweet types. Is a Preservative.
11	Monocalcium phosphate	Improver	Also called acid calcium phosphate. Used as a raising agent, and in bread. To be deleted from list of improvers

and biscuit flours. Self-raising flour, which is intended for cake baking, usually contains less protein than plain. Bread flours tend to be made from imported wheat, though new baking techniques have enabled more British wheat flour to be used, see Bread.

Self-raising flour, which accounts for about 80% of retail sales, is usually white, containing permitted additives and **raising agents**. Government regulations specify a minimum amount of gas (carbon dioxide) which must be produced when self-raising flour is cooked. It should be kept dry and used as soon as possible after buying because its potency declines during storage, particularly in damp conditions.

Cakes, biscuits and pastry (flour confectionery) are made with flours of 70% extraction or lower, fat, sugar, and a wide range of food **additives**, including **emulsifiers, colours, flavours, preservatives** and, carried in the fat, **antioxidants**. Flour confectionery is usually low in its content of essential nutrients compared to **energy** and should be avoided by those needing to lose weight.

FLUORINE—see also Teeth

A **trace element**. Like other trace elements, fluorine may be an essential **nutrient** in very minute quantities, but is toxic in excess. In small quantities, fluorine appears to increase the resistance of **teeth** to decay—other trace elements, for instance **molybdenum** and **vanadium**, may also have an effect. Free fluorine is a toxic gas (like **chlorine**) but in nature fluorine is in the form of salts, or fluorides. Fluorine in food and drink is usually called fluoride.

All diets contain some fluoride. Most foods contain traces—about 0·02 milligrams in each 100 grams (3½ oz) of food, or 0·2 parts per million (ppm). Sea fish contain more—between 5 and 15 ppm—but tea is the only significant source in the British diet. It contains up to 100 ppm dry weight, and a cup of tea supplies between 0·2 and 0·5 milligrams. The daily intake from food (including tea) is estimated to be between 0·6 and 1·8 milligrams, though heavy tea drinkers will consume more. Children consume the least because they drink little tea and eat a smaller total quantity of food.

Water contains variable amounts of fluoride. In Britain waters contain between nil and 6 parts per million of fluoride, adding between nil and 6 milligrams to the daily average intake. Most people drink about 1 litre (1¾ pints) of water each day (ppm = milligrams per litre). People who work in hot environments and others who drink more than a litre a day will consume more.

Fluoride is absorbed out of food and water into the blood stream: calcium decreases the amount that can be absorbed. Once in the blood stream fluoride is transported around the body and some is retained in the bones and—during the growth period before eruption—tooth enamel. The rest is eliminated from the body in urine and sweat. More is retained whilst bones and teeth are being hardened with calcium during growth, but little is otherwise certain of the fate of fluoride. It has recently been reported as an essential nutrient for animals, but not yet been shown to be essential for humans.

Fluoride is toxic in excess. A lethal acute dose for an adult is likely to be 2 grams, or about 1000 times the normal intake. Over the long term, chronic poisoning—fluorosis—has been reported in areas where the water supply contains over 10 ppm, and in workers in contact with fluoride-containing substances—for instance aluminium smelters. The pelvis, backbone and limb bones become increasingly dense, and the ligaments of the spine are hardened with calcium, resulting in an immobile back (poker back). In Britain, the only food which was likely to be contaminated with toxic quantities of fluoride was baking powder. The acid part (see Raising agents) used to be obtained from rock phosphates which also contain fluoride. These are no longer used, but present day regulations still specify maximum permissible quantities of fluoride in baking powder and self-raising flour—15 and 3 ppm respectively, or 1·5 and 3 milligrams per 100 grams.

In areas where the fluoride content of water is low, children usually have more decayed (or missing or filled) teeth than in those where the fluoride content of water is between 1 and 2 ppm (adding between 1 and 2 milligrams to the average diet). Fluoride probably becomes part of the chemical structure of enamel; when in contact with acids produced by bacteria as they multiply on the enamel, the fluoridated enamel is thought to dissolve less readily. Fluoride may also inhibit the growth of bacteria.

When the fluoride concentration of water is greater than 1·5 ppm, teeth become increasingly mottled. The enamel becomes opaque and there can be white specks (mild mottling) or yellow, brown or black markings (moderate to severe mottling). Mild mottling can also occur when there is virtually no fluoride in the water and is assumed to be idiopathic (having an unexplained cause). When water contains 1 ppm of fluoride, mottling occurs but is hardly detectable and similar to idiopathic mottling. At 2 ppm, about 10% of teeth are moderately mottled, and when the fluoride content of water is high (over 3 to 5 ppm) mottling of the teeth is common.

Water containing 1 ppm (adding 1 milligram to the average diet) of fluoride is considered to be ideal. At this level, there is least mottling and the incidence of decay is low.

Dental decay in children is a serious problem in most industrialised countries, and there have been many experiments to see if it can be reduced by adding fluoride at the optimum level to mains water supplies naturally low in fluoride. The reduction in decay claimed varies from 60 to 30%, depending on whether the results are compared with those of a 'control'—see below. Fluoride is usually added as sodium fluoride, or sodium silicofluoride or hydrofluosilicic acid—these separate out into fluorine, **sodium** and **silicon** particles once in the water. There is no difference between added and natural fluorine particles.

Since 1955, a Government study has been assessing the

effects of fluoridation: in three towns where the water content of fluoride was low, fluoride was added to the level of 1 ppm. The children in these towns were compared with those in three similar 'control' towns where the water content of fluoride—also low—was left unchanged.

The children in study and control towns were divided into groups, according to age. The amount of dental decay was assessed by counting the number of decayed, missing and filled teeth in individual mouths and calculating the average total for each group of children. This measurement cannot be used to assess decay in adults, where more teeth are lost as a result of gum disease. Consequently the effects of fluoridation after adolescence are not certain.

AVERAGE NUMBERS OF DECAYED, MISSING OR FILLED TEETH
PER CHILD *

| | | Average per child | | Percentage | |
		1956	Latest year	reduction %	Difference %
Age 3–7	Watford	4·8	1·8	63	33
	Sutton	4·1	2·8	30	
Age 8–10	Watford	3·0	1·5	50	31
	Sutton	3·0	2·5	19	
Age 11–14	Watford	6·4	4·3	33	19
	Sutton	6·3	5·5	14	

* *Reference* Rep. Pub. Health and Med. subj. no 122 HMSO

Between 1956 and 1967 the amount of dental decay in one fluoridated town (Watford) decreased by an average of between 50 and 60% for children aged 3 to 10. This supports the common statement 'Fluoride cuts dental decay by half'. However, the amount of decay in its control—unfluoridated Sutton—also decreased by an average of 20 to 30% over that period. A cut of 30%—or one-third—would be more accurate.

The average numbers of decayed, missing or filled teeth counted in these two towns are shown in the table above. In terms of average numbers of teeth, children who received fluoridated water for 8 to 10 years had an average of one tooth less decayed, missing or filled compared with those who did not. Those children in the fluoridated area aged 3 to 7 had an average of 1·7 less decayed missing or filled teeth compared with those in the control town. In the older age group (11 to 14) there was less improvement (or reduction in decay) probably because these children had not been given fluoridated water all their lives. Fluoride is more effective if it is taken whilst the dental enamel is being formed before teeth erupt. Once the teeth erupt the enamel is not in contact with the blood stream.

Taking control towns into account, there have been 28 and 58% reductions in fluoridated towns in Michigan and East Germany respectively for children aged 10. In one study in New York there was a reduction of 45%—and an increase in the control area of 15%.

In 1963, following 5 years of fluoride trials in Britain, the Minister of Health advised local authorities that they were free to add fluoride up to 1 ppm in waters that were naturally low in fluoride in order that the amount of dental decay could be reduced. However, in 1969, only 2 million people were drinking fluoridated water in Britain. Fluoridation has not been universally introduced because it has been vigorously opposed at local authority level. Unfortunately personal convictions have led to unsupportable claims being made by both sides.

There is no evidence of any relevant change in the health of people drinking water fluoridated to a level of 1 ppm, except that the amount of dental decay in children is reduced. The members of the Government Research Committee on Fluoridation have categorically stated that at a level of 1 ppm fluoridation is completely safe. On the available evidence, fluoridation does not increase the incidence of rheumatism, bone disease (see also Osteoporosis), cancer, gout, goitre, mongolism, and heart disease. However, it is uncertain (though unlikely in view of the absence of complaint from millions of tea drinkers) that fluoride is a cause of allergy. People treated with artificial kidney machines (which require large volumes of water for each treatment) could accumulate sufficient fluoride to cause fluorosis. Other people who need to drink large volumes of water (for instance steel workers and those suffering from some kidney stones (see Calcium)) may also be at risk, especially if they also consume a lot of strong tea.

Postulated health risks to adults apart, fluoridation opponents challenge the statistical methods used in fluoridation studies and maintain that the actual benefits for the individual are small. Even if fluoride is beneficial for children's teeth, opponents feel that adding fluoride to water supplies is wasteful because only about 1% is actually drunk. It is feared that large quantities of fluoride which would be released will eventually result in unacceptably high levels in the diet as a whole—especially where food is processed in fluoridated water.

The most important opposition argument is a moral one: fluoridation is a form of mass medication. In other nutritional improvement programmes, alternatives are available —for instance ordinary salt is available for those who do not wish to take extra iodine, and wholemeal flour is available for those who do not wish to take extra calcium and nicotinic acid in their bread. Although water fluoridation may reduce dental decay in childhood, its benefits in adult life are not certain. In adult life most teeth are lost as a result of gum disease, caused principally by poor eating habits and inadequate tooth brushing. Water fluoridation does not encourage children to take care of their teeth over the long term and has been called a 'technological fix'.

There are other methods of increasing the fluoride content of tooth enamel which are almost or as effective, see below, but require positive action on the part of parents and children. Water fluoridation is aimed at those who do not have the opportunity or inclination to take positive action. In this respect it is an enlightened policy, but should not be allowed to obscure other—more basic—methods of decay prevention. At present dentists do not receive payment for time spent teaching people how to look after their teeth and neither is the subject taught in schools. Few local authorities employ community nutritionists. Part of the cost might usefully be raised by a higher tax on sweets.

Fluoride tablets, toothpaste and fluoride preparations painted on to teeth by dentists are the most effective alternatives to water fluoridation. Fluoride tablets are as effective, provided they are taken daily. Each contain 1 milligram of fluoride (as 2·2 milligrams of sodium fluoride). The recommended dose from birth to 2 years is half a tablet daily which may be dissolved in water or sucked before swallowing. From the age of 2, one tablet should be taken daily, until about 12 years of age when all the teeth have erupted. Like other medicines, large quantities of tablets should be kept out of the reach of children. After the teeth have erupted, fluoride tablets are probably of little benefit, though externally applied fluoride—as toothpaste or other preparations—is probably worthwhile even in adulthood.

Fluoride preparations applied topically to the teeth by dentists probably reduce dental decay in children by about 30%, provided they are carried out regularly (once to twice yearly). Unfortunately this treatment is not available under the National Health Service.

Most toothpastes contain monofluophosphate (MFP) which when compared with a control reduces dental decay by about 20%. MFP contains approximately 13% of fluoride, and 10 grams (1 inch) of a toothpaste containing 0·8% of MFP supplies approximately 1 milligram of fluoride. Concentrations are stated on toothpaste containers (0·8% = 800 milligrams per 100 grams of toothpaste). If toothpaste is swallowed, fluoride tablets should not be given in addition.

FOLIC ACID—Group names—folic acid, folate, folacin
                    —Obsolete names—vitamin B$_c$, vitamin M,
                      Wills factor, leucovorin factor
                    —Chemical name for synthetic form—
                      pteroylglutamic acid (PGA)

Folic acid is part of the **vitamin B complex**. An insufficiency
in the diet results in **anaemia**.

All foods, except fats, sugar and spirits, contain the
vitamin, but it is easily destroyed during cooking and processing. Liver, oysters and yeast **extracts** are probably the
richest sources; nuts, vegetables (especially dark green), and
wholegrain cereals are good sources. Meat, fish, eggs, milk,
refined cereals, and fruit contain small amounts.

Green vegetables contribute the most folic acid in normal
diets—and vegetarians tend to eat more than others—but
even 'poor' sources (for instance bread and fresh milk) can
contribute significant amounts. Less folic acid tends to be
eaten in winter and early spring than in summer, when
plenty of fresh vegetables and salads are available.

Folic acid is sensitive to heat, light, oxygen and readily
diffuses into cooking water. Although good sources when
eaten raw, green vegetables may lose all their folic acid
during cooking. The vitamin is protected by vitamin C:
foods cooked by methods which preserve vitamin C will
incur the minimum losses of folic acid. Folic acid is probably
fairly well retained in other cooked foods—weight-for-
weight grilled, fried and baked meat, fish and eggs may
contain as much or more than when raw (these foods usually
shrink and lose water when they are cooked, becoming more
concentrated). More is lost if meat or fish are boiled or
stewed, and probably little can be recovered in the cooking
liquor. If it has lost its vitamin C, folic acid will be destroyed
when **milk** is boiled. The losses incurred during processing
are unknown—frozen foods probably retain most of the
vitamin, but canned foods may lose all of it.

Adults probably require between 100 and 200 micrograms

SOURCES OF FOLIC ACID[1]

	Food	Approximate quantity supplying 100 micrograms of folic acid
Rich sources	Liver, dark green vegetables[a] (like spinach, broccoli, spring greens, watercress)	Less than 100 grams (3½ oz)
Good sources	Kidney, peanuts, walnuts, bread with added wheat germ, cauliflower[a]	150 to 350 grams (5 to 12 oz)
	Eggs, lettuce, mushrooms, tomatoes, oranges	350 to 500 grams (1 lb)
Moderate sources	Cheese, carrots, cucumber, potatoes, peas, bananas, grapefruit, oranges, dates	500 grams to 1 kilogram (2¼ lbs)
Poor sources	Meat, fish, white bread, milk, apples, peaches, pears, prunes	More than 1 kilogram

a – 90% can be lost when vegetables are cooked
[1] – based on figures for 'free' folate (ie excluding very complex forms) given by Hoppner et al. (1972) *Canad. Inst. Food Sc. and Tech. J.* 5 no 2; Hurdle et al. (1968) *Am. J. Clin. Nut.* 21 10; and Chanarin *Getting the Most Out of Food* Van den Burghs Ltd 1975

of folic acid daily (one microgram is equal to one-millionth of a gram). More is required during growth, and children require more in proportion to their weight than adults. An increased intake—of at least 100 micrograms a day—is particularly important in pregnancy. The table below shows portions of food which probably contain 100 micrograms of folic acid.

There are many different forms of the vitamin: folate and folic acid are used to describe the group as a whole. Folacin —another group name—is used particularly in the USA. Folates in food are usually complex and it is uncertain how much of the very complex forms are available to the body. The synthetic form of folic acid—PGA—is one of the most simple and potent forms but only small amounts of PGA occur naturally in food.

During **digestion** the vitamin is converted to its active form by cells lining the small intestine. It is transferred to the blood stream and carried to the liver, where it may be stored.

Well nourished people have sufficient liver stores for 3 to 4 months. When needed the vitamin is withdrawn from the liver and carried in the blood stream to cells where it fulfils a fundamental role: it is a coEnzyme required for the replication and renewal of cells. When there is insufficient, growth and replacement slow down and eventually cease. Cells in the blood, skin and digestive system—among the most rapidly renewed—are first affected by lack of folic acid, resulting in anaemia, and skin and digestive disorders.

Deficiency of folic acid is common during pregnancy: it is difficult to meet the growing child's needs for the vitamin from food alone—particularly when multiple births are expected. Initially folic acid is withdrawn from the mother's liver stores to make good the inadequacy, but this may be insufficient. The ensuing anaemia checks the baby's growth and puts the mother's life at risk. Extra **iron** is also required during pregnancy, and the two nutrients are usually routinely prescribed as a combined tablet. During breast feeding a nutritious diet (containing plenty of liver, green vegetables and fresh milk) will enable the mother to supply her child with sufficient folic acid. Extra synthetic PGA is rarely necessary.

Other groups of people at risk from folic acid deficiency are the very young and the elderly. Babies have become anaemic when fed artificially with goats' milk—a very poor source—and the elderly are often unable to prepare a diet containing fresh food. Yeast extract (for instance Marmite) —or even $\frac{1}{2}$ pint of milk (if it is not left out in the sun or boiled)—is a convenient way of improving an elderly person's diet.

Although common in developing countries, folic acid deficiency is rare in adults in Britain, except when the diet is very restricted. One reported case had been living on a diet consisting solely of boiled milk, another on tinned baby food (the vitamin is now sometimes added to tinned baby food). Deficiency also complicates underlying disease, like **malabsorption** and **alcoholism**, and may be caused by drugs used for epilepsy, cancer, malaria and occasionally by the contraceptive pill.

Excess folic acid is harmless—it is filtered out of the blood stream by the kidneys but large doses may apparently cure pernicious anaemia (see Vitamin B_{12}) but not the degeneration of the nervous system. For this reason the British Code of Advertising Practice forbids the advertisement of over-the-counter vitamin preparations that contain sufficient folic acid to mask pernicious anaemia.

FRUCTOSE—other name, fruit sugar
                chemical name, laevulose

A simple sugar (**monosaccharide**). For the same sweetness, just over half as much fructose as sucrose (ordinary sugar) needs to be used.

The only foods which contain more than traces of free fructose are **honey** and a few fruits, for instance apples and pears. Nearly all fructose is eaten in a combined form as sucrose, which is split during **digestion** to fructose and glucose. **Invert sugar** is also used in the food industry.

Pure fructose used to be very expensive because it was in such short supply. It used to be made from **inulin**. Now, however, it can be made from starch using **enzymes** and a series of chemical manipulations. It is often sold as 'natural fruit sugar'.

Fructose supplies the same amount of energy, weight-for-weight, as other sugars and is not suitable for weight-reducing diets. However, small amounts can be used in the liver for **energy** without the need for insulin (see Diabetes) and it is sometimes recommended for diabetics treated with insulin and diet. Fructose also increases the rate at which the body uses **alcohol**, though the effect varies with the individual.

Fructose should be used with care. Unlike glucose it is absorbed out of the body into the blood stream by simple diffusion and excess quantities can result in diahorrea. More importantly, it tends to raise the level of some fats in the blood stream (see Lipids) and may also precipitate attacks of **gout** in susceptible people.

Fructose may have to be excluded from the diet of some infants suffering from rare Inborn Errors of Metabolism (which see). In infancy ordinary milk—provided it is not sweetened with sucrose—can be given. After weaning, a low blood **glucose** and resulting fits and brain damage, can be avoided if no sucrose, fructose, fruit or invert sugar is taken.

FRUIT

Refreshing, delicately flavoured foods. They are generally thought of as the sweet-tasting parts of plants, though botanically many vegetables, nuts, and cereals are fruits.

Some fruits are good sources of **vitamin C, vitamin A** (as **carotene**) and fibre. In Britain, blackcurrants and rosehips are the richest fruit source of vitamin C. A 100 gram ($3\frac{1}{2}$ oz) portion of stewed blackcurrants contains an average of 5 times the adult daily **recommended intake** of vitamin C. Strawberries and citrus fruit are also good sources—an average 100 gram portion has between 2 and 3 times the adult daily recommended intake. Mandarins contain slightly less than oranges, lemons and grapefruit. Apples, pears, bananas and grapes contain comparatively little—weight-for-weight, less than a properly cooked portion of cabbage or cauliflower.

Yellow (not white) melons, peaches and apricots are the best fruit source of vitamin A, though they contain much less than carrots. There is little or no carotene in white fruits (for instance pears). Fruits with tough skins and pips have more fibre than tender fleshy fruits.

Fresh fruits are 80—90% water and consequently they are low in **energy**. A small (120 gram) apple only contains approximately 40 kilocalories compared with 100 kilocalories in 25 grams (nearly one ounce) of sugar. In weight-reducing diets they should be chosen in preference to other desserts. Most of the energy in fruits is in the form of sugars (carbohydrates)—chiefly **glucose** and **fructose** but also (especially in apples and pears) **sucrose**. Bananas and grapes contain more sugars than other fruits and consequently are higher in energy. Twenty-five grams of banana contains about 20 kilocalories, compared with 10 kilocalories in 25 grams of eating apple. Dried fruits contain much less water and correspondingly are much higher in energy—25 grams of raisins contain 60 kilocalories.

Fruit flesh contains only traces of fat—fruits are usually allowed liberally in low fat diets—and little protein (less than 1%). Fruits also contain small amounts of B vitamins (though not vitamin B_{12}), **minerals** and vitamin E, but other foods are better sources of these nutrients.

The refreshing quality of fruits is due partly to their high water content and also to weak, sharp-tasting acids. Vitamin C is an acid (**ascorbic acid**) and **citric acid** is in all citrus fruits, pineapple and soft summer fruits. Malic acid is in apples and plums; benzoic acid (also used as a **preservative**) in cranberries; and **oxalic acid** in strawberries and rhubarb. Although the acid taste of some fruits may be uncomfortable for those with digestive disorders (see Dyspepsia), fruits do not increase the acidity of the body— see Ash. As fruits ripen, their acid content falls. Unripe fruits also contain **starch** and **cellulose** which is converted to sugar as they ripen. See also Pectin.

In the average British diet, about 25 oz of fruit and fruit products are eaten per person per week (excluding meals taken outside the home) which supplied about one-third (34%) of the total vitamin C intake in 1973. Vegetables and potatoes supplied over half (52%). The most popular fruit is the apple—which may be a reflection of a well known adage. It has been said, however, that an apple a day for some people may keep the dentist very busy. An excess of acid fresh fruit and other foods may erode the dental enamel (see Teeth). Lemons and citrus fruit juices are the most acid.

The vitamin C content of fruits varies markedly according to species and variety. Apples for instance have been

AVERAGE NUTRIENTS IN 120 GRAMS (4½ OZ) PORTIONS OF SOME RAW FRESH FRUITS

Nutrient	Apple	Banana	Cherries	Blackcurrants	Gooseberries	Grapes	Grapefruit	Melon	Orange	Pear	Plums	Raspberries	Rhubarb	Strawberries	Tangerine
Energy (kilojoules)	225	390	235	145	185	315	110	115	175	200	190	125	30	130	170
(kilocalories)	55	90	55	35	45	75	25	25	40	50	45	30	7	30	40
Carbohydrate (grams)	15	25	15	10	10	20	5	5	10	15	10	5	1	5	10
Protein (grams)	¼	1¼	¼	1	¼	¼	½	1	1	¼	½	1	¼	¾	1
Potassium (milligrams)	145	420	330	450	205	340	280	325	235	150	225	270	510	195	185
Magnesium (milligrams)	5	50	10	20	10	5	10	20	15	10	10	25	15	15	15
Calcium (milligrams)	5	10	20	70	20	15	20	20	50	10	15	50	125	25	50
Iron (milligrams)	0·35	0·5	0·45	1·5	0·7	0·4	0·3	0·6	0·4	0·25	0·4	1·45	0·5	0·85	0·3
Other minerals	Small amounts														
Vitamin C (milligrams)	5	12	5	240	50	5	50	30	60	5	5	30	12	70	35
Vitamin A (microgram equivalents)	6	40	25	40	35	0	0	Yellow 400	10	2	45	15	12	6	20
Vitamin E (milligrams α tocopherol)	0·6	0·2	—	1·2	0·5	—	0·4	—	0·2	—	0·8	—	—	0·2	—
B vitamins: folic acid (micrograms)	2·5	18	—	—	—	—	12	—	25	6	—	—	—	—	—
others	Small amounts	Small amounts (no vitamin B₁₂)													

— = no figures available

AVERAGE NUTRIENTS IN SOME DRIED AND CANNED FRUITS

Nutrient	In 50 grams (2 oz) dried fruit (uncooked)							In 120 grams canned fruit and juice			
	Currants	Dates	Figs	Prunes	Raisins	Sultanas	Apricots	Mandarins	Pears	Peaches	Pineapple
Energy (kilojoules)	510	520	450	340	520	525	305	325	320	335	320
(kilocalories)	120	125	110	80	125	125	75	75	75	80	75
Carbohydrate (grams)	30	30	25	20	30	30	20	20	20	20	20
Protein (grams)	¾	1	1¾	1¼	¼	¾	¼	¼	¼	¼	¼
Potassium (milligrams)	350	375	500	430	430	425	305	120	110	180	70
Iron (milligrams)	0·9	0·8	2·1	1·5	0·8	0·9	0·8	0·2	2·1	2·3	2
Calcium (milligrams)	50	35	140	20	30	25	15	10	5	5	20
Other minerals	Small amounts	Small amounts					Small amounts	Small amounts			
Vitamin A (microgram equivalents)	—	4	4	85	0	0	200	—	2	50	8
Vitamin C (milligrams)	0	0	0	0	0	0	6	—	1	5	10
B vitamins: thiamin (milligrams)	0·02	0·04	0·05	0·05	0·05	0·05	0·02	—	0·01	0·01	0·09
others	Small amounts	—	—	—	—	—	—	—	—	—	—
Vitamin E	—	—	—	—	—	—	—	—	—	—	—

— = no figures available

reported to contain between 2 and 30 milligrams of vitamin
C per 100 grams, and oranges between 35 and 80 milligrams
per 100 grams. The values given in the table are only a
guide.

Freshly picked fruits contain more vitamin C than stored
because the vitamin declines slowly during storage. Its
destruction is hastened by warmth, bruising and drying out.
In apples, pears and bananas (which are poor sources of
vitamin C) these losses are not important, provided some
vegetables or citrus fruit are taken each day. Citrus fruits
however should be kept as cool as possible, preferably in a
refrigerator.

During cooking, the acidity of fruits protects against
severe loss of vitamin C, and compared with vegetables,
little is lost. Up to 90% is normally retained, but prolonged
soaking, cooking and keeping hot (especially if fruit is
puréed) can destroy all vitamin C in apples.

Some **additives** are permitted on some fresh fruits, which
should be washed or peeled before eating. Apples and pears
may contain **antioxidant** to prevent browning of the flesh
during storage, and bananas and citrus fruits may have
added **preservative** to prevent mould growth during storage.
Citrus fruit is also allowed to be coated with **mineral hydrocarbons** to prevent drying out during storage.

Fruits are preserved by drying, treatment with sugar (a
form of dehydration), **wine** making, and bottling, canning
and freezing.

Dried and crystallised fruits are concentrated sources of
sugar and should be avoided by weight reducers and
diabetics. Dried peaches, prunes and apricots retain 25 to
50% of their vitamin A, but dried fruits do not contain
vitamin C. Dried and crystallised fruit is permitted to contain
the presevative sulphur dioxide (which helps to protect
vitamin A) and glacé fruit usually contains added **colour**.
Crystallised fruit is made by dipping in hot sugar solutions,
which destroys most of the vitamin C, but candied peel
retains a little. The peel of citrus fruits contains more

vitamin C than the flesh. See also Preserves (jam and marmalade).

During bottling and canning heat destroys up to 60% of vitamin C: stored bottled fruit will lose more if exposed to sunlight. Blanching before freezing results in some loss of vitamin C, but the losses are smaller than those which would be caused by enzyme action during storage.

Canned or bottled citrus fruit juices (but not squashes—see Soft drinks) are very reliable sources of vitamin C. They are acid and 80 to 90% of vitamin C is retained. Little is lost during storage but vitamin C in contact with oxygen rapidly declines: up to 50% can be lost from an open can in a week. Only small quantities should be opened and kept under refrigeration. Blackcurrant syrup is also a reliable source of vitamin C, but apple juice contains much less than citrus fruits. Fruit juices are permitted to contain the preservative sulphur dioxide (which aids retention of vitamin C) or benzoic acid.

Small amounts of fruit juices are used to make fruit squashes and cordials. The comminuted pulp may also be used—the drinks are called 'whole fruit'. Unless vitamin C is added, squashes contain little or no vitamin C—see Soft drinks. The skin, left over from canning, freezing and soft drink industries is a source of essential oils. These are expressed (squeezed out) and used as flavourings in food and perfume industries.

GALACTOSE

A simple sugar (or monosaccharide).

The majority of galactose in the diet is derived from lactose (milk sugar) which is split during digestion to its component sugars—glucose and galactose. These sugars are absorbed into the blood stream, and galactose is normally converted to glucose in the liver.

Galactosaemia is an Inborn Error of Metabolism where the normal conversion of galactose to glucose does not take place. The accumulation of a toxic intermediate compound causes damage to many tissues, including the brain, liver and lens of the eye. Blindness and mental retardation can be avoided if galactose is excluded from the diet in time.

Special lactose-free milks are available on prescription for galactosaemic babies. After weaning, normal foods can be eaten, provided all galactose containing foods are avoided. These include milk and all milk products (for instance cheese and cheese spreads, yoghurt) and offals like liver, brains and sweetbreads. Lactose and milk are also used in many processed foods (some breads, cakes, biscuits, soups, tinned meats) and **labels** must be read carefully. Lactose is sometimes added as a 'filler' in tablets. **Agar, carragheen, raffinose** (in beetroot and molasses), and stachyose (in **pulses**) are also usually forbidden: although these carbohydrates are not usually digested, they may be fermented by bacteria in the large bowel and galactose absorbed into the blood stream.

GELATINE (and jelly)

A **protein** of limited nutritive value: it is lacking in the essential **amino acid** tryptophan and contains little phenylalanine.

When **meat** is cooked in water, collagen—the main protein in connective tissue responsible for toughness of meat—is converted to gelatine. Bones and gristle—which contain more connective tissue than meat—yield more gelatine. On cooling, gelatine causes the cooking liquor to set—'richer' liquors contain more gelatine and form firmer jellies. Calves' foot is a good source.

Commercially, gelatine is produced by simmering hide, skin and bones in water. It is added to many meat products and is the basis of sweet jellies. Home made jellies may contain vitamin C if they are made from **fruit** juices, but proprietary products—mixtures of sugars (sucrose, invert sugar, glucose), gelatine, flavouring and colour—do not contain fruit. Apart from **energy** and a small amount of poorly used protein, they are an insignificant source of **nutrients**.

NUTRIENTS IN A 150 MILLILITRE ($\frac{1}{4}$ PINT)
PORTION OF JELLY, MADE WITH JELLY CUBES

Energy (kilojoules)	515
(kilocalories)	120
Protein (grams)	3
Carbohydrate (grams)	30
Minerals	Insignificant amounts
Vitamins	0

GLUCOSE—other names, dextrose, grape sugar, blood sugar

A simple sugar (or **monosaccharide**) which can be made in the body from many food constituents, including **protein**. Only a few natural foods—for instance **honey** and grapes—contain more than traces of free glucose, but it is found in abundance as combined forms—mostly as **starch** (a polymer of glucose) and as **sucrose** or sugar—a **disaccharide**. Commercially, glucose and glucose syrups are made from starch using acids or **enzymes**.

Despite claims for certain glucose drinks and sweets, glucose has no particular **energy** giving properties. Although it may be absorbed into the blood stream at a faster rate than starch, weight-for-weight it contains the same amount of energy as other **carbohydrates** (approximately 4 kilo-calories per gram). However, glucose has only half the sweetness of sugar (sucrose) and is sometimes useful for special high energy diets: for the same sweetness, twice as much glucose can be taken.

After **digestion** glucose, from sugars and starches, is absorbed into the blood stream and transported to the liver. Some is immediately circulated round the rest of the body, and some is converted into **glycogen**. The blood glucose level never falls to zero—the brain and nerves need a constant supply for continued activity—and between meals it is maintained by glucose released from liver glycogen. The liver is able to replenish its own glycogen reserves with glucose manufactured from protein, **galactose**, some **fructose**, and other substances released from cells.

This complex system is under the control of hormones and enzymes. The malfunctioning of one link can cause a variety of symptoms—see Diabetes.

In cells, glucose is degraded in about twenty stages and its energy is harnessed for energy requiring reactions (see Metabolism). The last series of reactions is called the **citric acid** cycle. Finally, oxygen is combined with hydrogen: waste products—carbon dioxide and **water**—diffuse into the blood stream and are eliminated from the body. Carbon dioxide is breathed out through the lungs; and the kidneys, skin and lungs remove excess water.

GLUTAMIC ACID

An **amino acid**. It is a component of food **proteins**, the vitamin folic acid, and **monosodium glutamate**, but is not an essential **nutrient**. It is readily synthesised by the body and incorporated into new proteins, required for replacement and growth, plays an important part in the formation of other non-essential amino acids. Excess intakes are converted to **glucose** and used for energy purposes. The nitrogen part is used to make other amino acids or converted to urea, later filtered out of the blood stream by the kidneys.

Monosodium glutamate is made from glutamic acid extracted from sugar beet and wheat gluten.

GLUTEN

A **protein** capable of stretching when mixed with water and kneaded. Present in wheat, rye and, to a lesser extent, barley and oats. It is essential for leavened bread and cake making, when it expands and traps gas (see Raising agent) to form a spongy cooked product.

When flour is stored, gluten becomes tougher and more

elastic, resulting in a larger loaf. Flour improvers (see Flour) accelerate this process which would otherwise take several weeks. Gluten is added to starch reduced and high protein breads. They are lighter (contain more air) but their energy value is the same or greater than ordinary bread.

Strong flours—for instance Canadian—contain more gluten and are imported to make the British loaf. However, weak flours are best for cake and biscuit making. New methods of bread making allow more British flour to be used—see Bread.

A gluten-free diet is used in the treatment of coeliac disease. Special gluten-free flours and bread are available on prescription for those who need them: the difficulties encountered in baking with gluten-free flour can partially be overcome by adding protein in the form of milk or eggs.

Gluten-free diets are of dubious value in the treatment of multiple sclerosis, see also linoleic acid. About one-third of all cases are mild, allowing a normal lifespan and job, with little disablement. In more serious cases there are long remissions. These cases (or mistaken diagnoses) may account for the cures claimed to result from a gluten-free diet.

GLYCEROL—other name, glycerine

A sweet substance, part of the structure of fats (see Fatty acids), and made in the body. Excess intakes are used for energy by the same series of reactions which degrade glucose.

Glycerol is manufactured from fats and used as a humectant—particularly in cakes—see Additives, and as a solvent.

GLYCINE

The simplest amino acid. Because of its sweet taste, it is sometimes used in conjunction with saccharin. Glycine is contained in all food proteins, but is not an essential nutrient. Excess intakes are converted to glucose and used for energy purposes. The **nitrogen** part is used to make other amino acids or converted to urea, which is later filtered out of the blood stream by the kidneys.

Glycine is used to make new body proteins needed for growth and repair and for other important substances, including bile. Potentially toxic substances, for instance benzoic acid (see Preservatives), combine with glycine in the liver and are later eliminated in the urine.

GLYCOGEN—other name, animal starch

A polymer of **glucose (polysaccharide)**, used as an **energy** reserve in muscle and liver cells. Normally the energy stores of glycogen are small—most people contain less than 1 kilogram of glycogen, compared with up to 16 kilograms of fat in a 70-kilogram (11-stone) man.

Glucose released from liver glycogen stores maintains the blood glucose level, ensuring a continual supply of energy to the brain and nerves between meals. In muscles, glycogen is an emergency store of energy, independent of nutrients supplied in the blood. For sudden activities, like running, glycogen is broken down and a small amount of energy obtained by the conversion of glucose to **lactic acid**, without the need for oxygen. Modern athletic training regimes usually incorporate a high **carbohydrate** diet which increases muscle glycogen stores and probably enhances endurance during long distance events. The reserves are further increased if they are first depleted by several days' hard exercise and a carbohydrate-free diet.

Molluscs, such as oysters, which are cooked alive, contain a little glycogen. Other fish and meat are carbohydrate free.

GOUT

Arthritis, caused by deposition of uric acid crystals in joints. There is a marked familial tendency to gout, and it is most common in middle-aged men.

Uric acid is a waste product, continually made in the body and filtered out of the blood stream by the kidneys. Its precursors—purines—are components of the nucleic acids (DNA and RNA) which form the genetic code in cells. Nearly all foods contain some purine, but the body also manufactures it from **amino acids** using **folic acid** (which is a coEnzyme).

In many people (about 3 in every 100 in Britain), uric acid tends to accumulate resulting in a raised level in the blood stream. Few of those with a raised level of uric acid in the blood stream—only about 3 in every 1000 in Britain—actually develop gout. The disorder is exacerbated by a rich diet and excessive alcohol. Occasionally gout is a side effect of other medical conditions.

A typical first attack manifests itself as an excruciating pain in the big toe. This is the joint most subjected to stress (from shoes), but other joints—though not the spine—can be affected. The joint is swollen and tender for a few days, but eventually the pain passes off. There is usually a long remission, but if not treated the attacks—accompanied by fever, misery, irritability, and disinclination to eat—become more frequent. In severe gout, lumps of uric acid crystals (tophi) appear in the skin.

Gout is treated by drugs, but diet is important in reducing the frequency of attacks. The majority of gout sufferers are overweight. A well balanced **slimming diet** can reduce the blood level of uric acid, and, with a reduction in weight, the stress on foot and leg joints is minimised.

Once weight has been reduced, the diet can be normal, but heavy meals, excessive **fat** and—when medically advised—foods rich in purines (see table) and large helpings of meat should be avoided. All alcoholic drinks raise the blood uric acid level: in severe gout alcohol may have to be omitted entirely. Fasting and excessive **fructose** are two other factors which can precipitate an attack and high fat, carbohydrate-free diets are particularly harmful. Tea and coffee consumption can be normal: **caffeine** (although it is a purine) is not converted to uric acid.

Uric acid crystals tend to precipitate in the urine of gout sufferers and people with a high blood uric acid level, increasing the risk of kidney stones and possible kidney failure. The urine should be kept dilute: at least 4 pints of

FOODS RICH IN PURINES

Offals	sweetbreads
	liver
	kidney
	heart
Yeast	and yeast extracts
Roes	cod roe
	caviar
	herring roe
Fish	anchovies
	whitebait
	sprats
	herrings
	mackerel
	salmon
	sardines
	mussels
	scallops
Meat	partridge
	guinea fowl
	meat extracts (for instance Bovril)
	(other meats contain moderate quantities)

non-alcoholic beverages—and more in hot working conditions—should be taken. See Slimming diets for low energy drinks list.

Gout is to some extent inherited. Three-quarters of sufferers have a family history and one-quarter of near male relatives have a raised blood uric acid. It is probably an advantage for male relatives of gout sufferers to seek medical advice because a high blood uric acid is probably a risk factor in **heart disease**. Women, whose hormones exert a protective effect, are rarely affected by gout before the change of life.

GRAVY—and sauces

May be an opportunity to recover **minerals** and some water-soluble **vitamins** leached out of food cooked in **water** or stock.

Heat-sensitive water-soluble vitamins (like vitamins C and B_1 (thiamin)) are largely destroyed in the cooking liquor, especially if cooking is prolonged. Heat-stable B vitamins and minerals are recoverable: in boiled or simmered meat and fish up to half the **riboflavin** (vitamin B_2) and nearly all the **nicotinic acid** leached out remains in the liquor.

HEALTH FOODS

Strictly, any wholesome food containing a balanced proportion of essential **nutrients** is a health food, but it is generally accepted that health foods mean those sold in health-food shops. They include 'organic', 'natural' and so-called miracle foods. Health-food shops are also outlets for vegetarian foods and various pills and tonics.

Organic foods are grown in soils fertilised with manure or compost, without the use of chemical fertilisers or pesticides. They are claimed to taste better and be more nutritious than ordinary foods. Natural foods are usually described as having 'nothing added or taken away', that is containing no artificial **additives** and unrefined. They are sometimes called 'wholefoods' or unprocessed foods. 'Miracle' foods are promoted by writers in health-food literature: some, like **yeast** and **wheat** germ are rich in minerals and vitamins. Others, like **honey**, are promoted for their supposed medicinal properties.

Organic farming may improve the structure of the soil, and is probably less harmful to the environment than current conventional methods. Foods produced organically are more expensive—yields are reduced and more of the crop is lost to pests—but enthusiasts feel the cost is worthwhile to avoid agricultural chemicals. Those who cannot

afford to choose are at no apparent risk from the present
level of pesticides (see Poisons in food) and may in fact
sometimes benefit. Fungicides prevent the growth of
moulds, the toxins of some of which are suspected carcinogens. Intakes of pesticides can be further reduced by avoiding
large helpings of animal fats, and by washing or throwing
away the outside leaves and skins of fruits and vegetables.
It should be remembered that organic food is not always
pollutant free: fall-out, and polycyclic hydrocarbons from
tractor exhausts, can be deposited on both organic and
ordinary food.

Whilst there is no evidence that organic foods are
significantly better nutritionally than ordinary (breed and
maturity influence vitamin and mineral contents more than
the quality of the soil), some health-food writers have
influenced food choice for the better by publicising the
possible harmful effects of eating too much convenience and
refined food. Although additives are probably safe at normal
levels of intake, it is possible to exceed acceptable limits by
replacing fresh food entirely with convenience foods. The
effect of larger intakes cannot be predicted. Some wholefoods, like wholegrain cereals—which, when taken as part
of a balanced diet boost intakes of B vitamins, vitamin E
and minerals—are more nutritious than their refined
counterparts. Wholegrain cereals also contain more fibre
which may protect against some diseases of the large
bowel.

On the other hand, health-food writers tend to condemn
all food processing out of hand. It has been pointed out that
natural foods are not necessarily 'safer' or 'purer' (see for
instance Plant poisons in food) and that it is sometimes an
advantage to have things 'taken away', like the pathogens
in raw milk and the cyanide in tapioca. The 'natural'
function of most food is not that it shall be eaten by man,
but that it should grow, reproduce, die and decay. Some
processing is essential for the health of consumers, and the
argument of extremists that all food should be eaten whole

can become impracticable. Few people would enjoy eating the shells of nuts, straw or bones.

The health food movement has also been criticised for claims made in its literature for 'miracle foods' (alleged to improve vitality and prolong life); cures and systems (like slimming diets based on cider **vinegar**); and mineral and vitamin supplements. Recommendations for vitamins and minerals include vitamins A and B_2 for cataracts; vitamin E for impotence; magnesium for epileptic fits; calcium for bad temper; and vitamins A and D for asthma and acne. Unfortunately these and like claims have little basis in fact. Diseases effectively treated by extra vitamins and minerals are discussed individually, or in the relevant nutrient entry. See also Therapeutic diets.

HEART DISEASE

Ischaemic heart disease (IHD) is the commonest cause of death in Great Britain. It is almost invaríably caused by coronary artery disease, brought about by **atherosclerosis**. Other terms that mean or imply IHD are coronary heart disease (CHD), coronary thrombosis, heart attack, myocardial infarction.

The heart is a hollow muscle, nourished by **nutrients** and oxygen carried in the blood stream which circulates through a network of coronary arteries and veins. The coronary arteries are very susceptible to atherosclerosis: they become partially obstructed by fatty and fibrous material, and the muscle below the obstruction becomes deprived of oxygen (ischaemia). The resulting pain (angina) occurs especially during exercise and passes off after a rest. Eventually the obstruction—or, more frequently, a blood clot (thrombus)—may occlude the artery completely. Unless blood from another artery is diverted to the ischmaemic area, the cells below the obstruction die (myocardial infarction). Large areas of dead muscle can stop the heart altogether, but smaller areas are eventually healed by scar tissue. Minor heart attacks may pass unnoticed. Scar tissue cannot contract, consequently the extensively scarred heart is unable to cope with undue exertion.

IHD is—as yet—rare in developing countries, but has reached epidemic proportions in some affluent countries. However, there are several important differences in its incidence—in Finland for instance about three times more people per 100,000 die of IHD than in Sweden. The incidence is much greater in Scotland and Northern Ireland than in England and Wales. It is likely that IHD is caused by a combination of several risk factors. The most well established are shown in the table overleaf.

Risk factors in heart disease

Though it is not a proven cause of atherosclerosis (it may be a secondary effect) a raised level of fat in the blood is a major risk factor. Two important constituents of blood fat (strictly **lipid**) are **cholesterol** and triglycerides (simple fats).

Blood cholesterol is raised when the diet is rich in cholesterol and some saturated **fatty acids** (found in animal **fats**); blood triglycerides are raised when the diet is very

RISK FACTORS IN HEART DISEASE

		Notes
Not related to nutrition	Cigarette smoking	—
	A sedentary life	—
	Probably stress	
	Male sex	Six times more common in the 35–55 age group in men than women. Probably female hormones exert a protective effect during child-bearing years. Sex difference disappears in *diabetes*.
	Increasing age	
Related to nutrition	High blood pressure	May be caused by a high *salt* intake. May be a complication of *obesity*.
	High level of blood fats	May be caused by underlying disease, including inherited disorders (see Lipid) diabetes, obesity, kidney disease
	High level of uric acid in blood stream	See Gout
	Living in a soft water area	See Water

rich in sugar (both white and brown). The effect of sugar is more marked when the diet is also high in saturated fatty acids. Polyunsaturated fatty acids lower the blood cholesterol and triglycerides. They are mainly found in plant and fish oils, but there are exceptions—see Oils and Margarine. Polyunsaturated fats are as 'fattening' as other fats and they are less effective in lowering the blood cholesterol than saturated fats are in raising it.

The benefits of substituting polyunsaturated fats for saturated fats in the diet remain uncertain, but a reduced consumption of *all* fats and sugar is beneficial. Blood lipids are lowered and the likelihood of obesity reduced. Fried foods, and other fatty foods (see Fat) should be avoided, together with sugar in drinks (including soft drinks). **Fructose** and **sorbitol** should not be taken instead of sugar. People who are not overweight, or who lead a very active life, can safely increase consumption of starch (from bread, potatoes and other cereals). These changes are best initiated in childhood, since atherosclerosis is a progressive disease.

Except under medical advice, an actively blood cholesterol lowering diet (see Cholesterol) is probably of limited advantage. Increased regular mild exercise, like walking, and avoidance of smoking, stress and over-eating are more important. Sudden strenuous exercise should be avoided by unfit persons (including young men) otherwise a heart attack might be precipitated.

Other dietary factors which have been implicated in IHD include a lack of dietary **fibre**; an imbalance of trace elements in the diet; too little pyridoxine; and a high consumption of coffee.

Most cholesterol is eliminated from the body as bile salts. **Pectin**, a component of dietary fibre, has been shown to lower blood cholesterol, perhaps by combining with bile salts in the digestive system, thus hastening the elimination of cholesterol from the body. However, doses far greater than could be obtained from food were used; for instance,

3 kilograms (6½ lb) of apples would have to be eaten daily
to obtain an equivalent amount. Cereal fibre, as **bran**, had
no effect, even in equally large doses.

Trace element contents of foods are altered during
processing, for instance during the milling of wheat. Too
little is presently known of the relevance of trace elements to
heart disease, and whether modern diets are lacking, but
substitution of wholemeal bread for white is beneficial
because the fibre content of the diet is increased.

Diets deficient in **pyridoxine** (vitamin B_6) cause atherosclerosis in monkeys, though the implications of this finding
for human heart disease are unclear. Wholemeal bread
contains four times more pyridoxine than white. More than
6 cups of coffee taken regularly each day was found to
increase the risk of IHD in an American study, but in other
studies, the risk was due to accompanied cigarette smoking.

HEAT in foods

All other things being equal, hot foods contain slightly more
energy than cold ones. When hot foods (approximate
temperature 60°C) are eaten, they cool to normal body
temperature (37°C), transferring their heat to the body.
Heat is measured in kilo**calories**.

For instance, a hot mug of tea cools down 23°C (60°—37°),
and has a volume of about ½ pint, or 300 mls. Since 1 litre of
water releases 1 kilocalorie when it cools by 1°C (see
Calories), the energy given off by the tea is $\dfrac{23 \times 300}{1000}$, which
is equal to 7 kilocalories.

HERBS

Usually leaves of plants containing aromatic oils. Small
quantities added during cooking enhance and impart **flavour**
to savoury foods, stimulating appetite.

Herbs have a similar nutritive value to other green

vegetables, but as such small amounts are needed their contribution to the average intake of **nutrients** is usually negligible. Parsley however, can be used more generously: 2 teaspoons (2 grams), freshly chopped, sprinkled on to a portion of food contains an average of 3 milligrams of vitamin C and 70 microgram equivalents of vitamin A— about one-tenth of the adult daily **recommended intakes** for these nutrients.

Vitamin C rapidly declines after chopping and pounding. Dried herbs that have lost their colour will also have lost their vitamin A and C.

HISTIDINE

A semi-essential **amino acid.** Children do not grow if it is absent from their diet, but adults can probably synthesise enough for their daily needs. It is found in all food proteins and amounts eaten in excess of needs are converted to **glucose** and used for **energy** purposes. The **nitrogen** part is converted to urea, later filtered out of the blood stream by the kidneys.

Like all other amino acids, histidine is needed to make new body proteins for growth and repair. It is also the precursor of histamine, a substance normally present in small amounts in cells. When allergens (see Allergy) enter the tissues it is liberated in larger quantities and is responsible for nettle rash. In severe allergies, massive histamine release may cause a sharp fall in blood pressure (anaphylactic shock).

HONEY

Mainly composed of **carbohydrate** (sugars) and **water.** On average, 10 grams (1 teaspoon) of honey contains about 4 grams of **fructose,** 3 grams of **glucose,** $1\frac{1}{2}$ grams of water, and 1 gram of **sucrose.** It also has small quantities of **dextrins, maltose,** wax, and **protein.** There is pollen in comb honey, but it is usually strained off from blended varieties. **Colour and flavour** in different honeys are due to pigments and flavours from the nectar of flowers visited by bees.

Honey is sweeter than sugar (sucrose), because it contains fructose, and lower in energy, because of its water content. For the same sweetness, honey supplies less energy and it is therefore preferable to sugar for people living sedentary lives.

Sweetness apart, honey contains only minimal quantities of **minerals** and **vitamins** and compared with other foods has no special nutritional benefits. Claims for medicinal properties of honey (and pollen) are without any scientific foundation, but in the absence of genuine research cannot be upheld or disproved.

AVERAGE NUTRIENTS IN 10 GRAMS HONEY

Energy (kilojoules)	120
(kilocalories)	30
Protein (grams)	0·05
Carbohydrate (grams)	7½
Fat	Trace
Potassium (milligrams)	5
Phosphorus (milligrams)	2½
Other minerals (milligrams)	Less than 1
B vitamins:	
riboflavin (milligrams)	0·005
nicotinic acid (milligram equivalents)	0·02
others	0 or traces only
Vitamins C and A	0

HYDROLYSED PROTEIN

Animal or vegetable protein partially split to its constituent **amino acids**. It is used to give a meaty **flavour** to food, or as a flavour enhancer, often with **monosodium glutamate.**

ICE CREAM

A mixture of milk, fat, sugar, **emulsifiers, flavours and colour**. Except for dairy ice-cream, commercial varieties do not contain cream. Legally, ice cream must have a minimum of 5% fat, which is usually hardened (see Margarine) vegetable **oil**, containing **antioxidants**. Most ice creams have between 8 and 12% fat—soft ice-cream rather less (7%). In dairy ice-cream, all the fat must be from milk.

Ice cream has little nutritive value. About two-thirds of its weight is water, and half its volume air. Approximate nutrient contents per 60 grams (2 oz) serving are shown in the table below. Values for dairy ice-cream are similar, except that dairy ice-cream contains approximately 65 microgram equivalents of vitamin A. Non-dairy ice-cream contains none. Ice cream may have to be avoided in slimming diets and blood **cholesterol** lowering regimes.

AVERAGE NUTRIENTS IN 60 GRAMS ICE CREAM
CONTAINING 10% FAT

Energy (kilojoules)	495
(kilocalories)	115
Protein (grams)	2½
Fat (grams)	6½
Cholesterol (milligrams)	25
Carbohydrate (grams)	12
Calcium (milligrams)	80
Phosphorus (milligrams)	60
Potassium (milligrams)	100
Iron (milligrams)	0·15
B vitamins:	
thiamin (milligrams)	0·03
riboflavin (milligrams)	0·12
nicotinic acid (milligram equivalents)	0·66
others	—
Vitamins A, D, C	0

— – no figures available

INBORN ERRORS OF METABOLISM

Very rare inherited defects affecting the synthesis of a specific **protein**—usually an **enzyme**—in the body. Over two hundred different defects are known, and most result in disease early in life. Some involving an enzyme are treated with special diets—which may need to be synthetic. See also Malabsorption.

In **metabolism**, complex body constituents are built up and broken down by specific sequences of chemical reactions (metabolic pathways). Each chemical reaction is controlled by an enzyme. If an enzyme is lacking (or ineffective) in a metabolic pathway, substances produced before it in the chain accumulate in the body; later products are missing. Depending on the enzyme, the defect may pass unnoticed or lead to damage of the brain, liver and other vital organs. Damage can either be caused by toxic intermediates or by deficiency of nutrients at the end of the pathway.

Some defects respond to massive doses of **vitamins**— particularly those in the B complex, which are co**Enzymes**. In others, damage can be avoided by giving a diet low in substances at the beginning of the metabolic pathway. Phenylketonuria and galactosaemia—occurring in about one in every 15,000 and 70,000 infants respectively in Britain—are the two commonest inborn errors of metabolism successfully treated by diet. See Phenylalanine and Galactose. With prompt detection and treatment, mental retardation can be avoided. All infants are screened for phenylketonuria in the first few days of life, and although national screening is expensive, the cost is amply justified by the numbers of children who would otherwise be confined to mental institutions.

Gout, diabetes, and some blood fat (**lipid**) disorders have a familial tendency.

INOSITOL—obsolete names, Bois I, inosite, meat sugar

Like **paraamino benzoic acid**, and **choline**, inositol is some- times classed as a B complex vitamin. However, although it is required for the multiplication of yeasts and growth of some animals, inositol is not thought to be an essential **nutrient** (and is therefore not a vitamin) for humans. No human deficiency has ever been discovered, and sufficient quantities for daily needs are presumed to be made in the body.

All diets contain some inositol. It is part of **phytic acid** and good sources are wholegrain cereals and wholemeal bread. It is also part of **phospholipids** found in vegetables and animal protein foods—particularly meat. Most people eat about 1 gram a day—more than the daily needs for most of the known vitamins.

Mice fed an inositol-free diet lose their fur, but supplements of inositol do not prevent hair loss in humans.

INULIN

A polysaccharide (polymer) of **fructose**, found in many plants—for instance artichokes, endive and salsify. It used to be the commercial source of fructose. Untreated inulin has no nutritive value because it cannot be digested. Humans lack the necessary **enzymes** for the dismantling of inulin into its smaller components of fructose—see Digestion.

INVERT SUGAR

A type of syrup, extensively used in the food industry. It contains equal quantities of **glucose** and **fructose** and is made when **sucrose** (sugar) is heated or treated with acid or **enzymes**. It also occurs naturally, for instance in **honey**.

IODINE

An essential **trace element**, needed by the thyroid gland for synthesis of thyroid hormones which regulate many diverse and important body processes. Dietary insufficiency results in goitre.

Sea fish, seaweed and iodised salt are the only reliable sources of iodine. Sea salt contains a little, but it is lost during storage. When fish is boiled, up to half the iodine can be leached out into the cooking water, which should be used for sauce.

Other foods supply comparatively little iodine, and vary depending on the soil content of the area in which they are grown. The level of iodine in water is an indication of the iodine content of local diets, but drinking water in itself contributes little.

The daily needs for iodine are uncertain, but adults may require an average of about 100 micrograms (1 microgram is one-millionth of a gram). During pregnancy, breast feeding, infancy, childhood, and adolescence, about 150 micrograms may be required to ensure adequate supplies of thyroid hormones during growth. These needs are unlikely to be met in most areas in Britain unless one or two sea-fish meals are eaten each week, or iodised salt is used. The table opposite shows amounts of foods supplying an average of 100 micrograms of iodine.

During **digestion**, iodine is absorbed out of food and carried in the blood stream to the thyroid gland, where it is extracted. The gland stores about half the body content of iodine and incorporates it into thyroid hormone. Throughout the day, thyroid hormone is secreted into the blood stream, taken up by cells, and inactivated after it has exerted its effect. Iodine is then released and most is filtered out of the blood stream by the kidneys and eliminated in the urine. If iodine losses are not replaced by iodine in food, the thyroid gland increases in size in an attempt to trap more iodine from the blood stream and maintain an adequate supply of thyroid hormone. The enlarged gland causes the neck to swell, a condition known as goitre.

Small goitres are not harmful, and may be a natural adaptation. If treated early by increasing the iodine content of the diet, the gland returns to its normal size of about 30 grams (1 oz). In severe untreated cases it becomes very large, interferes with breathing and has to be surgically removed. Goitre may also occur in two serious diseases only rarely

IODINE IN FOODS[1]

Reliable sources	portions containing an average of 100 micrograms
	Iodised table and iodised sea salt, 4 to 6½ grams (about half the normal intake)
	Fish liver oils, 20 grams (about 4 teaspoons)
	Sea and shell fish, 125 grams (4½ oz)
	Salmon and sea trout, 300 grams (10 oz)
	Spinach, 500 grams (1 lb)
Other foods	Seaweeds (like *Carragheen*, laver bread, kelp) contain at least as much as seafish, but no analyses are available
	Other foods on average contain less than 5 micrograms per 100 grams (3½ oz) portion (at least 2 kilograms would have to be eaten to obtain 100 micrograms), but vary according to the iodine content of local soils. For instance, eggs reportedly vary from 1 to 20 micrograms per 100 grams, milk from traces to 15 micrograms per 100 millilitres

[1] Mainly based on values given in *Iodine Contents of Foods*, Chilean Iodine Education Bureau 1952

caused by diet—thyrotoxicosis and myxoedema (respectively over- and under-activity of the thyroid gland).

Despite the formation of a goitre, if the mother's diet is lacking in iodine during pregnancy and breast feeding, she may be unable to maintain adequate supplies for the growing child. Most brain growth takes place during the last 3 months of pregnancy (but is not complete until the second year of life) and if there is insufficient thyroid hormone, the brain fails to develop adequately, resulting in deaf mutism and cretinism. Welfare vitamin tablets available in pregnancy and breast feeding also contain 100 micrograms of iodine and these complications are now rare in Britain.

World-wide, about 200 million people suffer from goitre. It is endemic where the soil content of iodine is low, as in Derbyshire, the Cotswolds and Mendips. In some areas, goitre is found despite a moderate iodine water content, and genetic and other factors are thought to be involved. Plants of the Brassicae family (for instance cabbage, mustard) contain goitrogens which can interfere with extraction of

iodine out of the blood stream, or inhibit the synthesis of
the hormone in the thyroid gland, causing goitre in animals.
In Britain, humans consume too little of these foods to be
affected, but goitrogens are thought to contribute to goitre
formation in other parts of the world. Other trace elements,
for instance cobalt and manganese, are antagonistic to
iodine. In excess they provoke goitre in animals, but their
effect in humans is not known.

IRON

A **mineral** needed for red blood cell formation. **Anaemia** due
to lack of iron in the diet is one of the commonest nutritional
disorders, particularly likely to affect women.

Only a small proportion of iron in food is absorbed into
the blood stream during **digestion**. The iron in some foods is
better retained than in others. The richest sources of well
absorbed iron are black sausage, liver, kidney and red meat.
Other meat, meat extracts, fatty fish (like sardines), molluscs
(like mussels), and soya beans are good sources. White fish
contains comparatively little, but its iron is well absorbed.

Some vegetable foods—wholegrain cereals, other pulses,
nuts, dark green vegetables, curry, cocoa—and eggs are
apparently good sources of iron, but other constituents (like
phytic acid) render most of the iron in these foods unavailable for absorption. Thus, although spinach contains
nearly as much iron as beef, at least ten times the weight of
spinach would have to be eaten for the same amount of iron
to be absorbed. White flour is fortified with iron, to overcome some of the losses in milling (see Bread) but recent
research has shown the added iron to be poorly absorbed.
The availability of iron in these foods is improved by
vitamin C. For instance, more iron will be absorbed from a
breakfast of egg and toast if a glass of citrus fruit juice is
taken at the same time. Meat itself also enhances absorption
from these foods.

Poor sources of iron are sugar, fats, milk, cheese, yoghurt

and fresh fruit. In other iron-rich foods—like fruit gums, liquorice, glacé cherries, dried fruits, molasses (black treacle) and some wines—the iron is taken up from tools and containers used in manufacture and storage. Cooking utensils such as steel knives, can also contribute to the iron in the diet. It is uncertain how much iron from these sources is normally absorbed.

Like other minerals, iron leaches out of food into cooking water, but it can be recovered if the liquor is used for sauce or gravy. The iron content of foods is otherwise little affected by cooking, storage and processing.

Recommended intakes for iron take into account its poor availability from food. The table opposite shows portions of well absorbed iron containing foods which will supply the daily recommended intake (12 milligrams) for women up to the age of about 50. Children and adolescents require more iron in proportion to their weight than adults to allow for growth and more is needed during pregnancy and breast feeding to ensure an adequate supply for the growing child. See Appendix. The recommended intakes are lower for men than women. In Britain, the average diet (excluding meals taken outside the home) contains about 12 milligrams of iron, of which one-third is supplied by meat and fish.

Adults contain 3 to 4 grams of iron (a gram is one thousand times a milligram) which is carefully conserved.

FOODS CONTAINING WELL-ABSORBED IRON

	Food	Portion that will supply an average of 12 milligrams iron
Rich sources	Cockles, liver, black sausage, pigeon, kidney, winkles, mussels, Bovril	50 to 100 grams (1½–3½ oz)
	Corned beef, heart, hare, pheasant	110 to 150 grams (4–5 oz)
	Soya beans, oysters, whelks, duck, beef	170 to 200 grams (6–8 oz)
Good sources	Lamb, sprats, sardines, turkey	280 to 360 grams (10–13 oz)
	Bacon, pork, ham, chicken, roe	430 to 520 grams (15–18 oz)
	Herring, rabbit	630 grams (22 oz)
Poor sources	Luncheon meat, white fish	Over 1 kilogram (2¼ lbs)

About one-third is stored—mostly in the liver—and nearly all the rest is held in haemoglobin, the red pigment in blood cells which transports oxygen from the lungs to all parts of the body. Although individual red cells are dismantled every four months, their iron is recycled in replacement red blood cells. A small but important quantity of iron is contained in **enzymes**—present in body cells—concerned with the release of energy from food. Provided there is no bleeding, the only iron lost is from iron in these enzymes as cells die and are sloughed off from the outside of the body (for instance from skin). A healthy man only has to absorb sufficient iron from his food to replace these losses.

Because their needs are so small, men rarely suffer from iron deficiency due to a poor diet. However, when new blood has to be made, the iron requirement is greatly increased. Ten millilitres (2 teaspoons) of blood contains about 5 milligrams of iron. These increased needs are partly compensated for by less wastage of iron in food: about 20% (three times that normally absorbed by healthy men) can be transferred into the blood stream when there is a shortage. If the diet is poor in meat and other well absorbed iron-containing foods this adaptation may not supply sufficient.

If the diet is lacking in iron, it is initially withdrawn from liver stores to match the needs for new blood formation. When these are exhausted, anaemia develops—the newly-formed red blood cells are small and contain less haemoglobin. Minor degrees of anaemia may not be noticed (especially if little exercise is taken) but eventually the anaemic person suffers from the lack of oxygen and becomes tired and breathless. In severe cases there may be difficulty in swallowing, and angina (see Heart disease). Severe anaemias are life-threatening.

Women are particularly at risk: even in affluent countries, about 10% are estimated to suffer from iron deficiency. During the child-bearing years they must absorb at least twice as much iron from their food as men to compensate for regular monthly losses of blood. The diet must contain

at least one portion of meat or other good source of iron a
day, and ideally liver, kidney or other iron-rich food should
be taken once a week. Iron tablets are necessary once
anaemia has developed and for women whose monthly
losses are too great for sufficient iron to be absorbed, even
from a good diet. Tablets are also advisable for vegetarian
women, unless a lot of curry or soya is eaten.

During pregnancy extra iron is required to reinforce iron
stores in preparation for blood losses at childbirth and to
ensure adequate supplies for the growing child. The needs
are unlikely to be met from food alone and iron tablets are
routinely supplied at antenatal clinics in the last three
months of pregnancy: anaemia is associated with a greater
risk of complications for both mother and child.

When available, iron is transferred from the mother in the
last stages of pregnancy. This store tides the infant over the
first few months of life of milk feeding (milk is a poor source
of iron). The child may become anaemic if the stores are
inadequate or if milk feeding is prolonged—but infants
should not be weaned before 4 months. The iron in breast
milk is better absorbed than in fresh cows' milk, but iron
and vitamin C is added to some powdered baby milks.

Children and adolescents with inadequate iron stores may
not be able to meet the needs for new blood during growth
spurts without becoming anaemic. The recommended
intakes for iron during adolescence are as great as those for
pregnant women. Needs for iron are not increased in the
elderly, but their diet is often poor. Corned beef and
sardines are particularly valuable sources of iron—they are
easy to chew and require no cooking.

Anaemia due to blood loss is likely after surgery and
accidents. An iron-rich diet should be given during convalescence. It can also occur following regular unnoticed
blood loss from the digestive system—for instance as a result
of ulcers, haemorrhoids and large intakes of aspirin—
particularly in the elderly. See also Malabsorption.

For most people, iron supplements are not harmful: if

there is no shortage, the iron is not absorbed. However, like other medicine, iron tablets should be kept out of reach of children. Rarely, an inherited defect may allow abnormally large quantities of iron to be absorbed and an accumulation in the body may cause liver failure if not treated (by removal of blood). Alcoholics are likely to accumulate too much: some cheap wines are very rich sources, and alcohol and a low protein diet may interfere with regulatory processes, allowing too much iron to be absorbed.

ISOLEUCINE

An essential **amino acid** needed in the daily diet for the formation of new body **proteins** for growth and repair. Children fail to grow and adults lose weight if it is lacking in the diet.

Isoleucine is not a limiting amino acid—it is well supplied in all animal protein containing foods (except gelatine) and is also contained in vegetable proteins. Excess intakes are converted to **fat** or **glucose** and used for **energy**. The **nitrogen** part is converted to urea, a relatively harmless substance later filtered out of the blood stream by the kidneys.

Maple Syrup Urine Disease is an **Inborn Error of Metabolism** in which brain damage and early death can be avoided by a diet low in isoleucine and two other essential amino acids, leucine and valine. A synthetic diet is used until weaning after which the diet is based on low protein foods (sugar, fats, protein-free flour, and some fruits). Small measured amounts of good quality protein (usually as milk) are allowed to fulfil the child's minimum daily needs for the three amino acids.

JOULE

Basic unit for measuring **energy**. Unlike the **calorie**, which is only a unit for energy in the form of heat, the joule can apply to all forms of energy—for instance, work, heat, electricity, chemical energy, nuclear power. 4·184 joules equal one calorie.

Like the calorie, the joule is too small for practical dietary use and kilojoules (1000 joules) are used. 4·184 kilojoules (kJ) is equal to one kilocalorie (kcal). In some circumstances, when the total energy of a day's diet is measured, megajoules (MJ) are used. One megajoule is one million joules (or one thousand kilojoules).

Scientific journals and food tables now contain energy values expressed in joules. In other books, calories will be gradually replaced by joules. This changeover, like metrication is a result of the acceptance by Britain of the International System of Units (SI units) in 1960.

Some conversion factors are shown in the table below.

One kilocalorie is equal to 4·184 kJ
One Btu is equal to 1·055 kJ
One hph is equal to 2·685 MJ
One kW-hour is equal to 3·6 MJ

One gram of fat contains approximately 37 kJ
One gram of carbohydrate contains approximately 16 kJ
One gram of protein contains approximately 17 kJ
One gram of alcohol contains approximately 29 kJ

KWASHIORKOR

One extreme of protein energy **malnutrition** (PEM) in children. The other extreme is **marasmus**.

Although the name originated in Ghana, kwashiorkor (literally: the illness the older child gets when another child is born) is widespread in most developing countries. It usually occurs in the second year of life when the child is weaned on to a diet containing too little **protein**. The shortage of protein retards growth and affects all parts of the body but energy needs are often met (in contrast to marasmus). The skin flakes and may ulcerate; the hair is sparse and sometimes loses its dark colour (turning red); and the muscles may be so wasted that the child cannot crawl. The blood stream contains insufficient protein, and water accumulates: the child has a puffy appearance, and is less resistant to infections (which may themselves partly cause the disease). The liver may become infiltrated with fat (which sometimes leads to liver failure) and the digestive system degenerates. The resulting diarrhoea exacerbates the shortage of nutrients.

Any child is at risk but kwashiorkor occurs typically in areas where the staple food is low in protein (for instance tapioca (cassava)). Severe cases may recover with hospital treatment. In less serious cases supplements of skimmed milk or other protein restore the child. Unless the mother takes milk supplements for the child or is taught how to feed the child with local protein containing foods (if available) recurrence is likely.

LABELLING OF FOOD

Ingredients in nearly all retail prepacked food must by law be declared on the label. There are certain exceptions (see table overleaf): apart from flour and sugar confectionery and complete meals, these foods are covered by other government regulations which specify their composition. Very small containers do not have to declare their ingredients if there is insufficient space.

All ingredients must be declared in decreasing order of weight (as used at the 'mixing bowl' stage) except in the case of mixed fruits or vegetables which may be listed alphabetically. Water does not have to be declared, and foods sold in vinegar, syrup or other liquor may have the food listed first (even if it weighs less than the liquor). Dried or dehydrated foods may list foods in order of their weight when reconstituted.

All **additives**, except miscellaneous ones and those not controlled by permitted lists—see Additives, must be listed on the label. They may be declared by their chemical name or their group name—for example, either 'sulphur dioxide' or 'permitted preservative' could be used on the label of a container of dehydrated vegetables. Additive contents of ingredients must also be declared, unless in the case of **preservatives** and **antioxidants**, the ingredient contains less than 5% of the permitted level. Flavours, whether natural or artificial, together with their solvents, can be described as 'flavourings'. When flavours are used to simulate a food ingredient (which is either absent or in too small a proportion to flavour the food) the word flavour must be prominent —for instance, chicken flavour, raspberry flavour.

Certain foods are claimed to contain **nutrients**: only claims for energy, protein, vitamins A, D, C, three B vitamins (thiamin, riboflavin, and nicotinic acid), calcium, iron and iodine are permitted, and the label must state how much of the nutrient each 100 grams (or 100 millilitres) of the food contains. In the case of energy, the food must be a 'significant' source. Claims for protein are only permitted if the food contains at least 3 grams of protein for every 100 kilo**calories** the food supplies (this is less than, for example, eggs). In addition to the government regulations, there is a Code of Practice (not law, but it is used as guidance in legal proceedings) which forbids claims and advertisements that a food contains one or more of the above minerals or vitamins, unless a day's normal intake of the food would supply one-sixth of the adult **recommended intake**. No food should be described as a rich or excellent source unless a day's normal

FOODS EXEMPT FROM DECLARING INGREDIENTS WHEN
PREPACKED FOR SALE BY RETAIL

Any meal
Biscuits, cakes, and other flour confectionery
Bread (but not breadcrumbs)
Butter
Cheese (but not cheese spread and processed cheese)
Sweets, chocolates and other sugar confectionery
Flour (including self-raising)
Fresh fruit and vegetables
Ice cream (including dairy ice cream, milk ice, Kosher ice)

intake of the food would supply one half of the adult recommended intake and no food may claim to cure or prevent a disease unless a day's normal intake would supply all the recommended intake.

No food can be claimed to have specific slimming properties, but a food can be claimed to aid weight reduction as part of a diet low in energy (kilocalories), provided the claim can be substantiated, and its energy content (per 100 grams or 100 millilitres) and ingredients are printed on the label.

LACTIC ACID

A natural **preservative** in food. Its acidity discourages the multiplication of putrefactive microorganisms.

Lactic acid is formed from **glycogen** in meat and game. It is also made by some bacteria as they multiply on sugars (for instance lactose, milk sugar). Cultures of these bacteria are deliberately added to milk in the making of cheese, yoghurt, other cultured milks and sour cream. Apart from its preservative action, the lactic acid produced clots some of the milk protein, thickening yoghurt and sour cream. Olives, and cabbage (sauerkraut) are also preserved by fermentation with lactic acid producing bacteria.

LACTOSE—milk sugar

A sugar with about one-fifth the sweetness of **sucrose** (sugar). It is a **disaccharide** made by the milk glands of mammals and only found naturally in milk and milk products. Human and cows' milk contain 7 and 4 grams of lactose respectively per 100 millilitres ($\frac{1}{2}$ glass). Sucrose or other sugars are added to cows' milk intended for baby feeding in an attempt to make it similar in **carbohydrate** content to human milk.

In infants and young children, lactose is split during **digestion** to its constituent sugars, **glucose** and **galactose**, by the **enzyme** lactase which is made by the cells lining the

surface of the small bowel. These simple sugars are absorbed
into the blood stream and ultimately used for energy.
Caucasians are unusual in retaining the capacity to produce
lactase after weaning—most races lose this ability and
cannot tolerate large quantities of milk in adulthood. Only a
small quantity of lactose can be digested and the remainder
is fermented in the large bowel, causing diarrhoea. Cheese
and some fermented milks (which contain less lactose) are
better tolerated.

The capacity to produce lactase may also be lost as a result
of coeliac disease, an Inborn Error of Metabolism, and
surgery of the small bowel. The resultant diarrhoea and
failure to thrive in infants is cured by special lactose-free
baby milks. A lactose-free diet is also needed in the treatment
of galactosaemia (see Galactose).

LECITHIN—other names, phosphatidyl choline

A phospholipid, composed of glycerol, two fatty acids,
phosphorus and choline—see individual entries. It can be
made in the body and is not an essential nutrient, but it is
found in most foods. Egg yolks are the richest food source
(up to 10% by weight).

Commercially, crude lecithin is a mixture of several
phospholipids obtained from soybeans and other oil seeds
(like peanuts). Pure lecithin can be extracted out: both pure
lecithin and the remaining phospholipids are important
emulsifiers in the food industry.

Whilst all lecithins have the same basic structure, lecithins
from different sources contain different types of fatty acids.
Lecithins are the main phospholipids (emulsifiers) in the
blood stream and are present in the myelin sheath of nerves,
yet, despite claims by some enthusiasts, extra lecithin is
unlikely to promote healthy nerves and offset the risk of
heart disease. The body makes enough lecithin to satisfy its
requirements from the raw materials methionine (an amino
acid converted into choline), sugars or fats (to make fatty

acids and glycerol) and phosphorus. Small amounts of **linoleic acid** are also required.

LEUCINE

An essential **amino acid** needed in the daily diet for the formation of new body **proteins** for growth and repair. Children fail to grow and adults lose weight if it is lacking in the diet.

Leucine is not a limiting amino acid—it is well supplied in all animal protein containing foods (except gelatine) and is also contained in vegetable proteins. Excess intakes are converted to fat and used for **energy**. The **nitrogen** part is converted to urea, a relatively harmless substance later filtered out of the blood stream by the kidneys.

Maple Syrup Urine Disease is an **Inborn Error of Metabolism**, treated with a diet low in isoleucine, leucine and valine.

LINOLEIC ACID—other name essential fatty acid
—obsolete name, vitamin F

An essential **nutrient** for children and probably adults, contained in all **fats**.

Linoleic acid is the initiator of a group of polyunsaturated **fatty acids**—the 'essential fatty acids'—with an arrangement of double bonds that cannot be imitated by the body. The group is involved in the formation of **phospholipids** and the structure and functioning of cells, particularly nervous tissue. Essential fatty acids are also precursors of the prostaglandins, but only about 1 milligram a day of linoleic acid (less than one ten-thousandth of the daily intake) is used for this purpose.

At least 13 prostaglandins have been identified in different parts of the body, including the eye, kidney, lungs, brain and nervous tissue, intestines and reproductive organs. Some are able to make the heart beat with greater force;

others lower blood pressure, stimulate the intestine muscles to contract, and inhibit the release of fat from adipose tissue (fat stores). Others are able to inhibit blood clotting. Very little is known about their precise role in health and disease, but they may affect the development of, or recovery from, a heart attack.

Safflower, sunflower, soya, cotton seed and corn oils are rich sources of linoleic acid—over half their fatty acids are linoleic. Peanut oil and chicken fat are good sources; olive oil and pig fat (bacon, lard, pork) are moderate (about 10% of their fatty acids are linoleic). Beef fat, milk fat (butter, cheese and cream), coconut and cocoa butter are poor (less than 5% of their fatty acids are linoleic). Small quantities of other essential fatty acids also occur in food—linoleic and arachidonic acids respectively in plant and animal fats.

The adult daily requirement for linoleic acid is thought to be from 3 to 6 grams. The average British diet contains 12 grams of polyunsaturated fatty acids (mostly linoleic), half of which is supplied by meat, fish, cheese, eggs, milk and butter, and about one-third by margarine and cooking fats and oils. High polyunsaturated fat diets supply up to 60 grams of linoleic acid daily: **vitamin E** requirements are greater in these circumstances.

An adequate supply of linoleic acid is essential for babies. It is incorporated in the insulating (myelin) sheath of nerves and brain, whose growth is not complete until two years of age. Breast milk contains about 4 times more linoleic acid than liquid cows' milk, and more if the mother consumes large quantities of vegetable oils. Some baby milks are skimmed before drying and the fat replaced with vegetable oils containing an equivalent amount of linoleic acid.

In adults, no disease resulting from a lack of linoleic acid has been established—probably because only small quantities supplied by even the poorest of diets, are necessary once growth has ceased. Linoleic acid deficiency has been proposed as a cause of **heart disease**. This is unlikely because the acid is deposited in the fatty accumulations of **athero-sclerosis**. When supplements are taken, the composition of adipose tissue (the body's energy reserve) changes. Abnormally high quantities of linoleic acid are accumulated and the fat becomes 'softer'. The implications of this change are unknown.

Multiple sclerosis is a condition in which the myelin sheath of nerves is lost, often causing disablement. Sufferers have abnormal levels of linoleic acid in their blood stream. In the initial stages (when the myelin sheath is being broken down) the levels are raised; afterwards they fall. It has been suggested that supplements of linoleic acid (from sunflower seed oil) would increase the chances of myelin regeneration. It is, however, very uncertain if supplements are able to ameliorate the disease: attempts to establish their value are in progress.

LIPID

A biochemical term, sometimes used to distinguish fatty substances in the body from fat in the diet. When referring to food, **fat** is used: nearly all (98%) of fat in the diet is in the form of simple fat (**triglycerides**).

In the body, lipid has two main functions: it is used for **energy** and it is part of the structure of all cells. The fat in food can either be used immediately for energy, or deposited in energy stores (adipose tissue). Between meals, most cells rely on **fatty acids** released from adipose tissue and carried in the blood stream for their supplies of energy. The brain and nerves are exceptional in requiring a supply of **glucose** for their continued activity.

Adipose tissue—congregations of cells swollen with triglyceride (simple fat)—is situated under the skin and around organs. It has important secondary functions, cushioning the vital organs against injury and insulating the body against heat loss. Men carry about 12% of their weight as fat, women about 25%. Depending on the amount of energy eaten in food and used up in daily activities, individuals vary greatly in the amount of fat they hold in store. The fat cells may become larger and (at least during growth) increase in number. A body containing over 20% in a man (14 kilograms or 2¼ stones for a 70-kilogram (11-stone) man) and 30% in a woman would be considered obese—see Obesity. Fat stores can be reduced to as little as one kilogram (2¼ lb) during **starvation**.

Whereas the amount of lipid in adipose tissue can vary from day to day, the quantity of lipid in the structural part is normally not altered. The structural body lipids are of great variety and complexity. Nearly all contain fatty acids and most contain phosphorus (**phospholipids**). Some contain **carbohydrate** (for instance galactose). In brain and nervous tissue, large complexes form the myelin sheath, which acts as an electrical insulator. The steroids—for instance **cholesterol**, sex hormones and cortisone—are also classed as lipids. All these materials can be made in the body and, with the exception of small amounts of **linoleic acid**, fat is not an essential **nutrient** (though diets very low in fat are unlikely to satisfy all other needs).

Triglycerides (simple fats) are not soluble in water and have to be emulsified (divided finely see Emulsifiers) before they can be transported in the blood stream. The emulsified particles are called lipoproteins, and contain triglyceride, protein, phospholipid and cholesterol. Raised levels of lipoprotein in the blood stream between meals is called hyperlipoproteinaemia: depending on the lipoprotein involved, blood cholesterol or triglycerides or both may be raised.

Hyperlipoproteinaemia is a risk factor in **heart disease**, and, like **gout** and **diabetes**, is partly inherited, partly caused by a faulty diet. The three disorders are likely to affect the same people and all are associated with **obesity**.

Alteration of the amount of fat in the diet, in an attempt to lower the level of lipid in the blood stream and offset the risk of heart disease, is particularly worthwhile for those suffering from hyperlipoproteinaemia. The diet may have to be augmented with drugs.

There are six types of hyperlipoproteinaemia, of which three are quite common. The exact disorder and individual treatment need to be medically decided. An outline of three of the diets used is given here, but specialised advice should be sought.

Type IV is the most common, affecting middle-aged men in particular. It also occurs in alcoholism, some types of kidney disease, obesity, diabetes, and sometimes during pregnancy and use of contraceptive pills. The blood cholesterol is normal, but the blood triglycerides are raised. A reduction in weight where necessary and curtailment of drinking are most important. Thereafter the diet is low in carbohydrate (less than 200 grams a day) and restricted in fats, sugar and cholesterol.

Type IIA is mostly inherited: there is a very high level of cholesterol in the blood stream, and it is treated with a diet very low in cholesterol and saturated fats. Polyunsaturated fats are taken in place of saturated fats (see Oils and Margarine) and **alcohol** may be taken in moderation.

Type IIB may be inherited or result from a rich diet. Both cholesterol and triglycerides are raised. Reduction in weight is the most important treatment, followed by a diet high in protein and restricted in cholesterol, carbohydrate, fat, sugar and alcohol.

LYSINE

An essential **amino acid** needed in the daily diet for the formation of new body **proteins** for growth and repair. When it is lacking children fail to grow, and adults lose weight, becoming nauseated, dizzy, and intolerant of noise.

Lysine is the limiting amino acid in **cereals**. Animal foods (meat, fish, cheese, eggs, milk) and pulses are, however, good sources—for instance the 9 grams of protein in 50 grams (under 2 oz) of cod supplies the estimated daily adult needs for lysine.

When eaten together, proteins supplement each other: good sources of lysine overcome the deficiences of cereals. Bread and cheese, rice and lentils, cornflakes and milk are good combinations of proteins—the mixed protein is efficiently converted to new body protein and (provided the daily needs for protein are not exceeded) little is wasted.

In the USA, lysine is added to bread in an attempt to increase the amount of wheat protein used by the body. Similar proposals have not been accepted in Britain because lysine is not the limiting amino acid in mixed diets—see Methionine.

Excess intakes of lysine are converted to fat and used for energy. The **nitrogen** part is converted to urea, a relatively harmless substance later filtered out of the blood stream by the kidneys.

MAGNESIUM

A **mineral** which has many fundamental roles in the body. It is widespread in foods and frank deficiency is known to occur only as a result of underlying disease, or in conjunction with other essential **nutrients**.

Magnesium is evenly distributed in foods and beverages, except white sugar, fats, spirits, some **convenience foods**, and most **soft drinks** which contain virtually none. Wholegrain cereals, nuts and spinach are apparently good sources (particularly bran and wheat germ products) but **phytic acid** (and **oxalic acid** in spinach) is known to interfere with the absorption of magnesium. It is uncertain how much of the magnesium in these foods actually reaches the blood stream: wholemeal bread contains less phytic acid than other wholegrain cereals.

Magnesium is little affected by processing and cooking of foods—apart from losses in milling (wholemeal flour contains at least four times more than white), and leaching into cooking water. Up to two-thirds can be lost when vegetables are boiled but it is recovered if the liquor is used for sauce or gravy. Fried, baked and grilled foods retain all their magnesium.

There is no British recommended intake for magnesium, due to lack of information, but adults probably require about 300 milligrams a day—an amount contained in the average British diet. The table facing shows portions of foods containing an average of 100 milligrams. In pregnancy and breast feeding, needs for magnesium may be greater: American recommended allowances are set at 450 milligrams daily. The additional 150 milligrams is supplied by, for example, 200 grams (5 large slices) of wholemeal bread in place of white.

During **digestion**, about one-third of the magnesium in a mixed diet is absorbed into the blood stream. Normally an equivalent amount is eliminated from the body by the kidneys, but in childhood and adolescence, more is retained (less is eliminated in the urine) to allow for growth. By adulthood, most people have accumulated about 25 grams (less than one ounce) of magnesium, of which two-thirds is held as salts in the skeleton. The majority of the remainder is a vital component of cells: it is an **electrolyte**, helps to maintain the structure of cells, and is a cofactor for many important **enzymes**. It also influences the utilisation of two other important minerals, calcium and potassium.

Prolonged diarrhoea, and **malabsorption** syndromes (for instance, coeliac disease) can deplete the body of magnesium. None may be absorbed from food, and in severe cases up to four times more magnesium than is eaten in food may be carried out of the body in the stools. Abnormal losses in urine—due to some kidney disease, excessive alcohol consumption and some drugs used to increase the flow of urine (diuretics)—can also induce magnesium deficiency.

Initially some of the magnesium stored in the bones can be withdrawn to overcome a temporary lack of magnesium in the diet, increased losses or inadequate absorption.

APPROXIMATE PORTIONS OF SOME FOODS SUPPLYING 100 MILLIGRAMS
OF MAGNESIUM

Foods	Approximate portion supplying 100 milligrams magnesium
Good sources	Wheat germ[a], bran[a], soya beans, almonds[a], brazil nuts[a], winkles, instant coffee powder	Less than 50 grams (2 oz)
	Wholegrain cereals[a] (oats, puffed wheat, wheat biscuits) shrimps, whelks, chocolate, *extracts*, malted milks, fruit gums, black treacle, peanuts[a], walnuts[a]	50 to 100 grams (2 to 3½ oz)
Moderate sources	Brown and wholemeal bread, crispbreads[a], cockles, whitebait, dates, figs, coconut[a], cobnuts[a], spinach[b]	100 to 200 grams (3½ to 7 oz)
	White bread, puffed rice, most meat, cheese and fish, bananas, blackberries, currants, raisins, sultanas, raspberries, baked beans[a], peas, fried potatoes, mustard and cress, chestnuts, liquorice	200 to 500 grams (7 oz to 1 lb)
Poor sources	Other biscuits, cornflakes, white rice, milk, eggs, roe, sausages, luncheon meat, most fruits, boiled potatoes, most vegetables, other confectionery, sugar, soft drinks, alcoholic drinks, honey, fats and oils	More than 500 grams (1 lb)

a – high in phytic acid
b – contains oxalic acid

Eventually, when the stores are exhausted, personality changes, irritation, lethargy, depression and fits ensue, and if not treated death may be precipitated by a heart attack.

It has been suggested that diets supplying marginal intakes of magnesium (less than 300 milligrams per day) may be a cause of **heart disease**. Diets of people living in unindustrialised countries (where the disease is comparatively rare) usually contain more magnesium. Additionally the magnesium content of the heart muscle of Americans who died from heart attacks has been found to be lower than in those who died from other causes. However the suggestion is not generally accepted: too little is known of the factors governing magnesium utilisation, and the finding may be an effect rather than a cause of the disease.

Excess magnesium is not harmful in itself because it is not absorbed into the blood stream, but magnesium salts (for instance Epsom salts) have a laxative effect, drawing water and other minerals out of the body. Excessive use, particularly in the elderly, can result in **potassium** depletion.

MAIZE—Indian corn

A **cereal** associated in some countries with the deficiency disease **pellagra**. Its nutritional properties differ slightly from other cereals: it is deficient in **tryptophan** the precursor of the vitamin **nicotinic acid**. Yellow (but not white) maize contains **carotene** (provitamin A).

Maize originated in South America but became established as a staple food in many poverty stricken areas. It is resistant to drought, gives a good yield and has a short growing season. Where the traditional method of steeping maize in lime before cooking was not followed, pellagra became endemic. Its nutritional shortcomings are not important in a mixed diet containing plenty of nicotinic acid or tryptophan (found in most other sources of protein).

Maize can be eaten whole, as popcorn. Like other cereals it is ground into flour: maize meal retains all the nutritive value of the whole grain, but corn hominy or grits—where the nutrients in the germ and outer husk have been removed —are equivalent to white wheat flour. The germ is a valued source of **oil**. Maize flour cannot be made into leavened bread (it contains no **gluten**) but is usually eaten as a kind of porridge (for example polenta in Italy) or flat bread (for instance Mexican tortillas).

Cornflour is extracted after steeping the grain in water. It is a useful thickening agent (and is the basis of custard

AVERAGE NUTRIENTS IN 25 GRAMS (1 OZ) OF SOME MAIZE FOODS

Nutrient	Cornflakes	Canned Sweetcorn	Popcorn	Maize meal	Corn grits	Cornflour
Energy (kilojoules)	385	100	400	375	370	370
(kilocalories)	90	25	95	90	90	90
Protein (grams)	1½	¾	3	2¼	2	Trace
Carbohydrate (grams)	20	5	20	20	20	25
Fat (grams)	¼	¼	1½	1	¼	¼
Sodium (milligrams)	260a	60a	1	—	—	15
Potassium (milligrams)	30	25	65	—	—	15
Calcium (milligrams)	2	1	4	3	2	4
Magnesium (milligrams)	4	5	—	—	—	2
Iron (milligrams)	0·3	0·1	0·7	0·6	0·5	0·35
Other minerals	Small amounts (maize meal and popcorn more than others)					
Vitamin A (microgram equivalents)	0	10	0	0b	0b	0
B vitamins						
thiamin (milligrams)	0·01c	0·01	0·09	0·08	0·01	0
riboflavin (milligrams)	0·01c	0·01	0·04	0·03	0·01	0
nicotinic acid (milligram equivalents)	0·2c	0·3	—	0·3	—	0
pyridoxine (milligrams)	—	0·06	—	—	—	0
pantothenic acid (milligrams)	—	0·07	—	—	—	0
biotin (micrograms)	—	0·8	—	—	—	0
folic acid (micrograms)	—	—	—	—	—	0
vitamin B_{12}	0	0	0	0	0	0
Vitamin C (milligrams)	0	1	0	0	0	0
Vitamin E (milligrams α tocopherol)	0·03	—	—	—	—	0

a – salt added during processing
b – 10 micrograms for yellow maize
c – if enriched, thiamin 0·27 milligrams, riboflavin 0·35 milligrams, nicotinic acid 4 milligram equivalents
— – no figures available

powder (see Eggs) and blancmange) but is a source of 'empty calories'. It is virtually pure starch.

Cornflakes were first marketed at the end of the ninteenth century by the Kellogg brothers. John Kellogg ran a naturopathy clinic. Had the existence of the vitamins been realised at that time, cornflakes would not have been promoted for their health-giving properties. Most of the B vitamins are destroyed in milling and toasting. Brand named cornflakes usually have added vitamins B_1 (thiamin), B_2

(riboflavin) and nicotinic acid, but unless specified on the packet, cheaper varieties (like supermarket brands) can be assumed to contain negligible amounts.

Sweetcorn is a different variety of maize, containing more water and sugar. It is eaten as a vegetable: its nicotinic acid is made available by ordinary cooking methods.

MALABSORPTION

Disorder of **digestion** resulting in the failure to absorb one or more **nutrients** into the blood stream. May be caused by lack of digestive **enzymes** (sometimes as a result of an **Inborn Error of Metabolism** or disease of the pancreas) or bile; parasites or other infections; **coeliac disease**; or following surgery of the digestive tract.

Failure to absorb fat interferes with the absorption of any vitamin or mineral—commonly iron, calcium and folic acid—which is carried out of the body in the fatty diarrhoea. Loss of weight, flatulence and general ill health may be accompanied by **anaemia**, **osteomalacia** and skin disorders. The underlying condition may have to be treated with a special diet, for instance a **gluten**-free diet. A low **fat** diet is sometimes necessary but aggravates weight loss (low fat diets are low in energy). Low fat diets are sometimes supplemented with special oils (MCT) which can be absorbed in the absence of bile or fat splitting enzymes.

Intolerance to **carbohydrate** commonly takes the form of **lactose** intolerance; inability to digest starch and other sugars occurs rarely. The undigested carbohydrate is fermented by bacteria in the large bowel, causing irritation. Food is hurried through the digestive system with resulting malabsorption of other nutrients. There is usually an acid diarrhoea. The disorders are treated by a diet lacking in the relevant carbohydrate.

MALNUTRITION

Ill health brought about by a failure to eat a **balanced diet**. One, or more often, several, **nutrients** may be lacking or in excess. It may also be caused by underlying disease—for instance **malabsorption**.

In affluent countries, most malnutrition is associated with too much food—see Obesity, Diabetes, Heart disease and Teeth. Deficiency diseases due to lack of essential nutrients alone are comparatively rare, but see Anaemia and Osteomalacia.

Deficiency diseases are more likely in both adults and children when poverty restricts both the amount and type of food available. They are common in most developing countries, where protein energy malnutrition (PEM) is the most serious public health problem, estimated to affect up to 40% (or 500 million children in 1975) under the age of five. It is caused by too little food or lack of protein or both—most cases are intermediate between the two extremes of **kwashiorkor** and **marasmus**. Any other essential nutrient may be lacking, see for instance vitamin A.

A child may recover from PEM, provided he is given food in sufficient time, and in mild cases growth and development are not affected over the long term. In severe cases, modern treatment has reduced the risk of death, but although the child may later catch up with his peers in terms of growth, the effect of PEM on subsequent mental ability is uncertain.

MALTOSE

A sugar (disaccharide) composed of two molecules of **glucose**. It is about 30% as sweet as sugar (sucrose), and is found naturally in small quantities in some plants and honey. Maltose is formed during the **digestion** of **starch** and is itself digested to glucose, which is absorbed into the blood stream.

Maltose is formed in sprouted wheat or barley. Malt

extract has a pleasant taste and is used in malted drinks, malt loaves, barley sugar and as a vehicle for cod liver oil, but has no special nutritive virtues. Malted barley gives **flavour** and **colour** to **beer**.

MANGANESE

A **trace element**, which is thought to be an essential **nutrient** for humans, though no disorder resulting from a lack of manganese has ever been established. It is required as a cofactor for several **enzymes**; in animals, synthetic diets deficient in manganese cause infertility and defects in the bones and nervous system.

Wholegrain cereals, nuts and tea are rich sources of manganese: one cup of tea supplies over half the assumed minimum daily needs. Some green vegetables (like spinach) are good sources. Refined cereals, meat, fish, eggs, fats, fruit, white sugar, milk and cheese are comparatively poor.

The manganese contents of diets vary between 2 and 9 milligrams per day, which is presumed to be sufficient. In excess (100 times greater than normally eaten), manganese may interfere with the absorption of iron in the diet, but it would be extremely unlikely to eat this amount in food. Toxic manganese gases, inhaled by miners of manganese ores, cause a disease called manganic madness.

MARASMUS

One extreme of protein energy **malnutrition**, widespread in developing countries. It is the child equivalent of **starvation**.

Initially, lack of food checks growth. If given sufficient food in time, the child is likely to catch up with his peers. Otherwise the child wastes away: fat stores are lost and used for energy; later, muscles are broken down. The child has a wizened appearance, the stomach may be swollen and diarrhoea develops because proteins are not available to renew the digestive system and its **enzymes**. The child is lethargic and less resistant to infections which may themselves be a partial cause of the disease.

Any child is at risk from marasmus but it typically occurs in the early months of life, often following early weaning from the breast. It is more likely in impoverished urban areas: the mother may stop breast feeding because she needs to work, or in response to advertising campaigns for artificial baby milks. The child may be given too little (the milks are expensive) and in insanitary conditions the feeds are likely to become contaminated. The child's illness will be worsened if, in an attempt to cure the resultant diarrhoea, the mother feeds the child on a water only diet.

MARGARINE

A flavoured and coloured emulsion of water in fat (see Emulsifiers). First patented as a cheap substitute for butter by Mège-Mouriès in 1869, it contained suet, skim milk, and minced cows udder. Modern margarines are sophisticated products of food technology, often eaten for health and convenience reasons rather than economic. The most expensive brands now cost as much as butter in Britain.

Margarine has almost the same nutritive value as **butter**, though it contains more **vitamin D**. Margarine provides about one-third of the total vitamin D in the average British diet. It should be chosen in preference to butter by those likely to need extra dietary supplies of vitamin D, for example, the elderly, and Asian immigrants. Unless butter is added, margarine contains no **cholesterol**. Polyunsaturated margarines (see below) should be chosen when a blood cholesterol lowering diet has been medically advised.

Virtually any source of **fat**, whether artificial, animal or vegetable, can be used to make margarine: refining enables the production of oils and fats free of taste, smell and colour. Most margarines are blends of vegetable or fish oils—the type used depending on world supplies. Different margarines for specific purposes (for example ice cream, baking, table

margarines) are made after modifying—particularly hydrogenating—and blending refined fats.

Oils are liquid at room temperature, due to their higher
content of unsaturated **fatty acids**. They can be made more
saturated, and harder (solid at room temperature), by
treatment with hydrogen and a nickel catalyst. Oils are never
completely hydrogenated and some of the double bonds
remain: hard baking margarines contain about the same
percentage of polyunsaturated fatty acids as butter (about
5%). Soft margarines contain about 30% of their fat as
polyunsaturates, but only those with polyunsaturated fatty
acid content of 50% or higher (like Flora) are recommended
for blood cholesterol lowering diets.

In the final stages, skim milk, salt, **flavours**, **colours**,
emulsifiers, and vitamins A and D are worked into the fat by
a complicated system of beating and cooling. Margarine is
also permitted to contain **antioxidants**.

Brands vary slightly in their composition, but by law
margarine must contain not more than 16% water and a
minimum of 80% fat. If butter is added, no more than 10%
of the total fat is allowed to be butter. All margarines must
contain between 760 and 940 i.us. of vitamin A per ounce,
and 80 to 100 i.us. of vitamin D per ounce. Average com-
positions of margarine are shown in the table below. Low
fat spreads (which contain too much water to be called
margarine) like Outline are valuable foods for slimmers: they
contain half the energy of margarine or butter weight-for-
weight (35 kilocalories per 10 grams), and the same amounts
of vitamins A and D.

AVERAGE COMPOSITION OF MARGARINE PER 10 GRAMS

Water	1½ grams
Salt	100 to 200 milligrams
Fat	8½ grams
Cholesterol	0 (2·5 milligrams when 10% butter added)
Energy	335 kilojoules (80 kilocalories)
Vitamin A	90 microgram equivalents
Vitamin D	0·8 micrograms

MEAT (including poultry, offal and game)

An important source of **protein, B vitamins, minerals** and **energy**. Although it is the most reliable source of iron, meat is not an essential item of the diet. Liver is a rich source of most **nutrients**.

The **fat** in meat influences the nutritional value of particular cuts considerably. Most of the vitamins, minerals and proteins are concentrated in the lean (which contains up to a quarter of its weight as protein), and most of the energy in the fat. The cost of a particular cut of meat is not a reflection of its nutritional value: lean stewing steak has a similar nutrient content to lean fillet steak.

All meat has at least 1 to 3% fat, containing **cholesterol, phospholipids** and unsaturated **fatty acids**. Visible fat (including marbling fat—deposits between the muscle fibres) is more saturated. Beef and lamb visible fat contains a higher percentage of saturated fatty acids than chicken or pork. The amount of visible fat can be as great as 60% (as in streaky bacon) but generally it is below 40%. Lean meats do not contain more than 20% fat. Generally older animals are fattier than young, the breast or belly is fattier than the leg, and pork is fattier than lamb or beef. Most poultry (except goose and duck), game and offal are lean.

In 1973, meat supplied about a quarter of the total **protein** in the average British diet (excluding meals taken outside the home). Meat protein is of good quality: when small quantities are eaten with foods supplying plenty of energy (like potatoes and fats) most of its protein is incorporated into new body protein. However, once the daily needs for protein are fulfilled, excess quantities are used for energy, and therefore wasted. Large quantities of meat (portions greater than 60 grams (2 oz) cooked weight) are unnecessary in most **balanced diets**.

Meat is a particularly good source of minerals **iron, potassium,** and **phosphorus**. In 1973 meat supplied one quarter of the iron in the average British diet (excluding meals taken outside the home). Red meats contain more than white, though it is well absorbed from both groups. Veal, rabbit, pork and chicken contain about half as much iron as beef and game. Liver, kidney, pigeons and black sausage are rich sources. Women of child-bearing age, whose needs for iron are greater than men, should include one portion of meat where possible in their daily diet.

Fresh meat (except liver and kidney) contains hardly any vitamin A and no vitamins D or C. It is a good source of all in the **vitamin B complex**, supplying about one-sixth of the total **thiamin** (vitamin B_1), one-fifth of **riboflavin** (vitamin B_2) and one-third of the **nicotinic acid** equivalents in the average British diet in 1973. Pork is richer in thiamin than any other meat. Meat is also a source of vitamin E.

Liver and kidney are of particular value in the diet because they are rich sources of iron, **vitamin A**, and all B complex vitamins (especially **folic acid** and vitamin B_{12}). They are sources of vitamins D and C. Despite their nutritional virtues, liver and kidney are not as popular as other meats— probably because they are easily overcooked and become toughened.

Fresh and frozen meat is not permitted to contain **colours, antioxidants** or **preservatives**, though mince may have added sulphur dioxide (a preservative) in Scotland throughout the summer months. Sulphur dioxide destroys thiamin. Tenderisers are permitted, but must be declared on an adjacent notice or, if the meat is prepacked, on the label. Tenderisers may be directly added to meat, or injected into the animal just prior to slaughter. Other additives are not specifically banned (except for vitamin C and nicotinic acid which preserve the colour of fresh meat and may mislead the customer) but their misuse could lead to prosecution under the Food and Drugs Act (see Additives). Oven-ready poultry is usually injected with **polyphosphates** and sometimes with butter or extracts of fat from older animals to improve the flavour. Meat may also contain traces of **antibiotics**. In sheep and cattle synthetic female hormones increase the lean in the

AVERAGE NUTRIENTS IN 60 GRAMS (2 OZ) OF SOME MEATS (VALUES FOR COOKED MEAT UNLESS OTHERWISE STATED)

Nutrient	Bacon (back, fried)	Roast beef (lean sirloin)	Roast chicken	Roast duck	Boiled ham (lean)	Lamb chop (lean, grilled)	Roast pheasant	Roast pork (lean loin)	Roast veal	Roast turkey
Energy (kilojoules)	1505[a]	565	475	790	550	685	535	715	585	495
(kilocalories)	360[a]	135	115	190	130	165	130	170	140	115
Protein (grams)	14	16	18	14	14	16	18	14	18	18
Fat (grams)	32[a]	7	4	14	8	10	6	12	7	5
Cholesterol (milligrams)	40	40	50	—	40	40	—	35	40	45
Iron (milligrams)	1·7	3·2	1·6	3·5	1·6	1·5	5	1·6	1·5	2·3
Potassium (milligrams)	310	215	215	190	270	240	250	210	260	220
Phosphorus (milligrams)	135	170	160	140	145	145	185	125	215	190
Magnesium (milligrams)	15	15	15	15	15	20	20	15	15	15
Calcium (milligrams)	5	5	10	10	10	10	30	5	10	25
Sodium (milligrams)	1670[b]	40	50	115	1260[b]	75	60	40	60	80
Other minerals	Small amounts									
Vitamins A, D	Trace	Trace	Trace	Trace	Trace	Trace	Trace	Trace	Trace	Trace
Vitamin C	c	0	0	0	c	0	0	0	0	0
B vitamins										
thiamin (milligrams)	0·24	0·05	0·03	—	0·3	0·06	—	0·48	0·04	0·02
riboflavin (milligrams)	0·09	0·15	0·08	—	0·12	0·15	—	0·12	0·16	0·07
nicotinic acid (milligram equivalents)	4·5	6·0	6·0	—	5·5	6·0	—	6·5	7·5	—
pyridoxine (milligrams)	0·2[d]	0·15[d]	0·64[d]	—	0·3[d]	0·2[d]	—	0·3[d]	0·2[d]	—

AVERAGE NUTRIENTS IN 60 GRAMS (2 OZ) OF SOME MEATS (VALUES FOR COOKED MEAT UNLESS OTHERWISE STATED)

Nutrient	Bacon (back, fried)	Roast beef (lean sirloin)	Roast chicken	Roast duck	Boiled ham (lean)	Lamb chop (lean, grilled)	Roast pheasant	Roast pork (lean loin)	Roast veal	Roast turkey
B vitamins										
pantothenic acid (milligrams)	0·2d	0·3d	0·5d	—	0·3d	0·3d	—	0·4d	0·4d	—
biotin (micrograms)	4d	2d	6d	—	3d	2d	—	2d	3d	—
folic acid (micrograms)	0·6d	2	2d	—	1·2	1·2d	—	1·2d	1·5	—
vitamin B_{12} (micrograms)	—	1·2d	—	—	0·6	1·2d	—	—	—	—
Vitamin E (α tocopherol milligrams)	0·2	—	0·1	—	—	0·4	—	—	—	—

a – lean grilled bacon will have similar values to lean ham
b – salt added in curing
c – 5 to 25 milligrams if vitamin C added in curing
d – values for raw meats
— – no figures available

AVERAGE NUTRIENTS IN 60 GRAMS (2 OZ) OF SOME CANNED MEATS, OFFALS AND SAUSAGES

Nutrient	Canned meats		Offals, uncooked			Offals, cooked		Sausages, uncooked	
	Corned beef	Luncheon meat	Heart (pork)	Kidney (lambs)	Liver (mixed)	Tongue (ox)	Sweet-breads	Pork	Black
Energy (kilojoules)	580	845	240	250	360	780	450	860	720
(kilocalories)	140	200	55	60	85	185	105	205	170
Protein (grams)	14	7	10	10	10	11	14	5	3
Fat (grams)	9	17	2	2	5	14	5	17	14
Cholesterol (milligrams)	—	—	90a	225	180b	—	150	—	—
Carbohydrate (grams)	0	3	0	0	0	0	0	6	9
Iron (milligrams)	5·9	0·7	2·9	7·0	8·3	1·8	1·0	1·5	11·7
Potassium (milligrams)	70	125	180	150	195	90	140	95	80
Phosphorus (milligrams)	70	100	45	150	190	135	355	65	15
Magnesium (milligrams)	15	5	10	10	15	10	10	5	10
Calcium (milligrams)	10	10	5	10	5	20	10	10	20
Sodium (milligrams)	830c	525c	50	150	50	1120c	40	460c	540c
Other minerals	Small amounts								
Vitamin A (micrograms)	Trace	Trace	35a	180	8780	Trace	—	Trace	—
Vitamin D (micrograms)	Trace	Trace	0	0	0·45	Trace	—	Trace	0
Vitamin C (milligrams)	0	d	—	7	18	0	—	d	0
Vitamin E (α tocopherol milligrams)	—	—	—	—	0·6	—	—	—	—
B vitamins									
thiamin (milligrams)	0	0·24	0·18	0·18	0·18	0·03	—	0·1	—
riboflavin (milligrams)	0·12	0·12	0·48	1·2	1·8	0·21	—	0·04	—
nicotinic acid (milligram equivalents)	4·6	3·7	5·2	7·6	10·3	5·7	—	2·3	—

| Nutrient | Canned meats | | Heart (pork) | Offals, uncooked | | Offals, cooked | | Sausages, uncooked | |
	Corned beef	Luncheon meat		Kidney (lambs)	Liver (mixed)	Tongue (ox)	Sweet-breads	Pork	Black
pyridoxine (milligrams)	—	—	0·12	0·4e	0·3	—	—	—	—
pantothenic acid (milligrams)	—	—	1·2	2·4e	4·4	—	—	—	—
folic acid (micrograms)	—	1·2	—	20	90–240	—	—	—	—
biotin (micrograms)	—	—	4·8	48e	56	—	—	—	—
vitamin B$_{12}$ (micrograms)	—	—	—	3·5	45	—	—	—	—

a – figure for beef heart
b – chicken liver 335 milligrams
c – salt added during processing
d – small amounts may remain if added in processing
e – figures for ox kidney
— – no figures available

carcass. They are either implanted as pellets in the ear or
given in the feed, but are discontinued before slaughter and
there is no detectable residue left in the meat. Hormone
pellets are also implanted in the neck of capons and turkeys:
they have the same effect as castration, a fatter bird with an
improved flavour, but a small residue is left in the meat.
Most (about 85 %) chickens are consumed as broilers, which
are not caponised (given hormones or castrated).

Cooking tenderises meat and also has a preservative
action, killing growing forms of putrefactive micro-organisms. However, cooked meat can easily be recontaminated
and cause food poisoning (see Poisons in food).

At least a quarter of the weight of meat is lost during
cooking: the protein shrinks, squeezing water out. Consequently, weight-for-weight, cooked meat contains more
protein than raw. Fat is also released, but cooked meat may
contain more than raw if it is fried or basted.

Vitamin and mineral losses during cooking are very
variable and depend on the type of cooking and cut of meat.
Generally minerals and two B vitamins (riboflavin and nicotinic acid) are all recovered in the cooking liquor. Thiamin
(vitamin B_1) is sensitive to heat: up to 70 % can be lost when
meat is boiled, stewed or braised and up to 50 % when
roasted or grilled. Up to 50 % of other B vitamins are also
lost.

When meat is cooked in the can, nutrient losses are of the
same order as those incurred during cooking. Corned beef
however, which is cooked in water before canning, contains
no thiamin and few other B vitamins. The liquor is used for
beef **extracts**.

Thiamin is well retained when meat is cured, and bacon
and ham are good sources. However, modern bacon and
ham contain less nutrients weight-for-weight than traditional
products. Traditionally, pork was cured by rubbing salt into
the carcass and smoking over wood fires, which resulted in a
loss of about 20 % of water. Now, brine solutions are injected (or the meat is soaked in tanks of brine) and smoke solu-

tions are painted on to the surface. Polyphosphates and
waterproof wrappers also aid water retention. The change is
most marked in wrapped sliced or tinned hams, which are
very moist. Bacon and ham may be an unsuspected source of
vitamin C, added during modern curing processes.

When meat is frozen, all the nutrients are retained but
thiamin decreases during storage. When held in a deep freeze
for six months, losses of between 20 and 40% of the vitamin
have been reported. The drip of thawed meats should not be
discarded: it contains up to 30% of the pantothenic acid and
10% of other B vitamins.

Meat products are permitted to contain most additives.
Antioxidants (apart from vitamins C and E or their synthetic
equivalents) may not be directly added but may be carried
over when used in an ingredient. All meat products contain
salt and most contain **monosodium glutamate**. Additionally
cured and pickled meats (like bacon, ham, luncheon meat,
tongue, corned beef) may have added **nitrite**. All canned
meats may be preserved with nisin. Sausages, and other meat
with cereal sold uncooked (like beefburgers, hamburgers),
may be preserved with sulphur dioxide.

The quantity of meat in nearly all meat products is controlled by government regulations: minimum specified meat
contents of some products are shown in the table facing.
Novel proteins may not be used to replace minimum meat
contents, although they may be used in addition. There are
no minimum meat contents specified for canned ham, gam-
mon and pork shoulder: these may contain a surprisingly
large amount of water in 'natural juices' set with gelatine.

Animals have an important role in farming: they convert
grass to a pleasant, nutritious and digestible food. However,
it is impossible to satisfy present demands for meat from
grazing alone and much of the world's food, for instance,
cereals, fish and soya concentrates, that could be eaten by
humans is given to animals. Although Britain produces sufficient food to feed all its population, paradoxically half of its
food is imported. Even allowing for modern advances (like

SOME EXAMPLES OF THE MINIMUM LEGAL REQUIREMENT FOR MEAT
IN MEAT PRODUCTS

Food	Minimum meat content % of contents	Comment
Canned meat alone	95 ⎫	
Minced meat alone	85	
Canned meat + jelly (like brawn), beefburgers, luncheon meat	80	60% of the minimum must be
Meat + gravy	75 ⎬	lean, i.e. a
Sliced meat + gravy, meat + sauce	60	maximum of 40%
Meat + gravy (or sauce) + stuffing + onion	40	fat
Curried meat	35	
Curried meat + rice	15 ⎭	
Faggots, meat balls, rissoles, pie fillings	35	
Pork sausages (80% must be pork)	65 ⎫	50% of the mini-
Others including beef and pork sausages	50 ⎬	mum meat content
Frankfurters and salami	75 ⎭	must be lean
Meat spreads and patés	70	
Potted meats	95	
Potted meats in jelly	70	
Meat pies	25 ⎫	Total weight of
Meat and vegetable pie	12½ ⎭	the pie
Spreads labelled 'requires grilling'	None	
Products labelled 'X with meat' (for example rice with meat)	None	
Pasties, meat with vegetable pies	None	

new breeds, the use of antibiotics), the conversion of plant
foods into animal flesh is very inefficient: only one-tenth of
grain is returned as edible food. Famines in the rest of the
world are aggravated (some would say caused) by the dis-
proportionately large quantity of the world's food supply
consumed by developed countries. Anyone aware of this
situation cannot justify large portions of meat. It should be
mixed with other foods: traditional recipes (pasties, pies,
meat with dumplings or yorkshire pudding, pasta and meat
sauce) are an example of the place of meat in a balanced
diet.

METABOLISM—see also Energy

Life is an unstable process of unceasing chemical changes—
metabolism—during which the energy released from the

breakdown of large molecules is used to make new molecules. Metabolism is derived from a Greek word meaning change.

The basal metabolic rate is a measure of the amount of energy used in self-maintenance—or the basic chemical changes necessary to maintain life. It includes the liberation of energy from **glucose** and **fatty acids** (see Lipid) reactions, together with energy-using reactions such as the synthesis of new **proteins** needed for growth and repair; the elimination of waste products from the blood stream by the kidneys; the maintenance of the body temperature and **electrolyte** concentrations in cells; and the contraction of muscles in breathing and in the heart to maintain a constant supply of **nutrients** and oxygen in the blood stream. Children have a higher basal metabolic rate because they are growing (synthesising new tissue); elderly people less than young adults because the number of active cells declines with increasing age.

Energy used in self-maintenance is about half the daily expenditure for most people. The remainder is used in movement: contraction of muscle is also a chemical process requiring energy. The body is only partially effective at converting one form of energy into another. Energy lost during both energy-using and energy-liberating reactions is dissipated as heat and used to maintain the body temperature.

The amount of energy used in self-maintenance, activity, and lost as heat is replenished by the energy released from food. Energy taken in excess of immediate needs is stored, mostly as fat—see Lipid.

METHIONINE

An essential **amino acid**—it cannot be made in the body and has to be obtained from **proteins** in food.

Cysteine is synthesised from methionine. It is also obtained from food proteins and reduces the daily needs for methionine: more is made available for incorporation into new body

proteins needed for growth and repair. Methionine is needed
for the synthesis of other important substances, including
choline. Cysteine, cystine, and methionine are the only
amino acids which contain sulphur; the three are grouped
together and called the sulphur amino acids.

The sulphur amino acids are usually the limiting amino
acids in diets, determining the quality of the protein—or how
much can be used for growth and repair. Although they are
the limiting amino acids in British diets, there is no shortage:
the estimated adult daily need is supplied, for example, by
the 25 grams of protein in 80 grams (3 oz) of roast chicken.
Excess intakes are used for energy purposes and the unwanted nitrogen is converted to urea. The kidney later filters
waste urea and sulphur out of the blood stream.

MILK

An important source of calcium, vitamin B_2, protein and
energy. Cows' milk is a substitute for breast milk in infancy
(see page 215), and a convenient source of the extra essential
nutrients needed in childhood, adolescence, pregnancy and
breast feeding. Like any other food, excessive quantities of
milk can lead to obesity.

Two-thirds of a pint of milk (enough for about 5 cups of
tea and a milk pudding or milk on breakfast cereal) supplies
the daily recommended intake of calcium for adults and
children up to the age of nine. One pint meets the recommended intake of calcium during breast feeding and pregnancy (see Appendix). In 1973, milk supplied about one-half
of the total calcium in the average British diet (excluding
meals and snacks taken outside the home). Milk contains all
other minerals. It is rich in phosphorus but a poor source of
iron.

Milk is a good source of riboflavin (vitamin B_2)—providing, in 1973, about one-third of the riboflavin in the average
diet—and contains all other B complex vitamins. There are
very small quantities of vitamin C in fresh milk.

The proteins in milk are of good quality: when eaten in small amounts nearly all are incorporated into new body proteins needed for growth and repair. One pint supplies more than the recommended additional intake of protein during pregnancy and breast feeding. Milk also contains **carbohydrate**—about 10 grams in each third of a pint—in the form of **lactose** (milk sugar).

Vitamins A, D and **E** are dissolved in droplets of **fat** dispersed throughout milk. Milk fat has a low content of polyunsaturated **fatty acids** and contains **cholesterol**. Summer milk contains more fat-soluble vitamins than winter, and Channel Island (gold top) milk more fat than others. Homogenised (red top) milk is treated to prevent the fat droplets coalescing and rising to form 'cream at the top of the milk'. The fat is removed from skimmed milk, together with the fat soluble vitamins and most of the cholesterol.

A pint of milk contains approximately 400 kilo**calories**— about the same amount of **energy** as 4 large slices of bread. Skimmed milk has half the energy value of full-cream milk, and is useful in weight reducing diets. Because it is lacking in fat soluble vitamins, skim milk is not suitable for baby feeding.

Milk from several mammals is used for food by humans. Docile herbivorous animals are preferred, like cows, goats, camels, sheep, horses and buffalo. The pig has never been used as a dairy animal because it can withold its milk until needed by the brood. Occasionally goats' milk is available in Britain but the majority sold is from cows. Goats' milk has a similar composition but contains less **folic acid. Anaemia** has sometimes occurred in babies fed boiled goats' milk.

Unless it is pasteurised, milk can be a dangerous carrier of micro-organisms, causing diseases like tuberculosis and scarlet fever. Unpasteurised milk is not permitted to be sold unless the herd is certified free of TB. Pasteurisation— heating the milk to above 61°C for 30 minutes—kills harmful bacteria, but a few harmless ones responsible for the eventual souring of milk remain. Pasteurisation causes minimal destruction of nutrients—see Table I. During delivery (milk is about one day old when delivered) vitamin C decreases further—to about 3 milligrams per pint.

Riboflavin (vitamin B_2) is not sensitive to heat and is unaffected by pasteurisation but it is very sensitive to ultraviolet light. When left outside for only 2 hours, half the riboflavin is destroyed in bright sun, and one fifth on a dull day. Milk sold in shops in clear bottles exposed to fluorescent lighting is likely to contain little riboflavin. Riboflavin is converted to a substance which in turn quickly destroys

NUTRIENT LOSSES IN PROCESSING AND STORAGE OF MILK—TABLE 1

Process	Nutrients lost	Comments
Pasteurisation	10% of the B vitamins thiamin (B_1), pyridoxine (B_6), folic acid, vitamin B_{12} 25% of vitamin C	
Boiling	25% of vitamin B_1 (thiamin). Slight loss of riboflavin in open pan	If all vitamin C lost, boiling destroys folic acid
UHT milk	Same as pasteurisation	50% of pyridoxine and vitamin B_{12} lost on 3 months' storage
Bottled	20% thiamin, B_{12}	
Sterilised	30% folic acid 60% vitamin C	
Evaporated milk	80% vitamin B_{12} 50% vitamin C and B_6 20% vitamin A 20% vitamin B_1	Vitamin D is added but evaporation is most destructive way of preserving milk
Condensed milk	10% vitamin B_1, B_6 25% vitamin C and folic acid 30% vitamin B_{12}	Less destructive than evaporation: added sugar reduces the need for heat to sterilise the can. Skimmed or full cream milk can be used.
Dried milk	Same as pasteurisation	Only significant loss is when fat soluble vitamins are removed from skimmed milk. Dried skimmed milk contains less than 1% fat. Whole (full cream) milk powder contains 26% fat. Partially skimmed (half cream) milk powder contains 14% fat.

any vitamin C. About one-quarter of thiamin and any remaining vitamin C is lost when milk is boiled. If all the vitamin C has already been lost, boiling destroys folic acid.

Milk is preserved by evaporation or drying (to remove water) and by sterilisation (heating above boiling point). Sterilised milk is packed in sterile bottles or cartons and keeps for several months without the need for refrigeration. Ultra-high temperature (Longlife) milk has as good a flavour and nutrient content as pasteurised, but some of its vitamins are lost during storage—see Table I for nutrient losses in preserved milks and during storage. Riboflavin is unaffected by preservation, but may be lost during storage unless milk is kept in opaque containers.

Vegetable oils can be added to dried skimmed milk to make filled milks. These have a higher content of poly-unsaturated fatty acids than normal milk, and little cholesterol. Non-dairy creamers are not a milk product. They are a mixture of fats, casein (milk protein) emulsifiers and other additives with hardly any nutritive value. See also Butter, Cheese and Yoghurt.

Milk is important in most welfare services. Table II shows the percentage of the recommended intake for some nutrients supplied by one pint of milk. In Britain, a daily pint is supplied free of charge to low income families and in other special cases. Although the elderly are not included in free milk schemes at least half a pint should be taken daily: milk is a cheap source of nutrients often lacking in an elderly person's diet.

APPROXIMATE PERCENTAGE OF RECOMMENDED INTAKE OF SOME NUTRIENTS SUPPLIED BY 600 MILLILITRES (1 PINT) MILK—TABLE II

Nutrient	Pregnant woman	Breast feeding woman	6-year-old child	16-year-old girl	35-year-old woman	65-year-old woman
Calcium	60	60	145	120	145	145
Riboflavin	55	50	100	65	70	70
Protein	35	30	45	35	35	40
Energy	15	15	15	15	15	20

Since 1946, all children were given a free $\frac{1}{3}$ pint of milk at school, but in 1971, supplies to children over the age of seven were discontinued except in special circumstances. The decision to abolish free school milk for all except 5- to 7-year-olds was viewed with concern by nutritionists: the provision of free milk was a demonstration of its importance in the diet of children and adolescents. For the same amount of energy, milk contains more other essential nutrients than most other foods. The alternatives to milk most likely to be available at break time—soft drinks, sweets and biscuits—are rich in sugar, a source of 'empty calories': the incidence of obesity, tooth decay and other forms of malnutrition could increase.

In convalescence, milk, especially dried, is a useful way of enriching puddings and soups. It often relieves the pain of peptic ulcers but should not be taken to the exclusion of other foods. Its low content of iron and (when boiled) vitamin C and folic acid, predispose to scurvy and anaemia. Most races are unable to digest lactose in large quantities of milk: cheese is better tolerated.

Baby feeding

Cows' milk—fresh, evaporated or dried—has been successfully used as a substitute for breast milk for millions of babies. Nevertheless, there are important differences in the nutrient of cows' and human milk (see Table overleaf). Some of the differences are overcome by diluting cows' milk —one part of water is mixed with two of fresh milk and a level teaspoon of sugar added for every 100 millilitres ($3\frac{1}{2}$ oz) of feed. The mixture must then be boiled to sterilise it. Instructions for the use of evaporated and sterilised milk are printed on the label of the tin and must be followed exactly. Welfare vitamin drops (containing vitamins A, D and C) are necessary for all babies, but the dose should be reduced for babies fed with powdered milks already fortified with vitamins.

When the mother's diet is adequate, breast feeding is pre-

AVERAGE NUTRIENTS IN SOME MILKS

Nutrient	in 200 mls (⅓ pint) whole fresh milk Cows' (pasteurised)	Breast (1)	in 25 grams (1 oz) preserved milks Dried skimmed	Condensed (full-cream)	Evaporated
Energy (kilojoules)	555	630	340	370	160
(kilocalories)	130	150	80	90	40
Protein (grams)	7	2	8½	2	2
Carbohydrate (grams)	10	14	12	14	3
Fat (grams)	7	9	Trace	3	2
Linoleic acid (percentage of total fatty acids)	2	10	—	2	2
Cholesterol (milligrams)	30	30	5	9	8
Calcium (milligrams)	240	60	315	85	75
Phosphorus (milligrams)	195	30	265	60	65
Potassium (milligrams)	320	100	335	100	125
Sodium (milligrams)	100	35	150	35	40
Magnesium (milligrams)	30	20	30	10	10
Iron (milligrams)	0·16	0·1	0·13	0·04	0·05
Other minerals	Small amounts	—	—	—	—
Vitamin A (microgram equivalents)	80	120	1	30	30
Vitamin D (micrograms)	0·06	b	0	0·02	0·02a
Vitamin C (milligrams)	1	b	2·5	0·8	0·4
Vitamin E (milligrams α tocopherol)	0·2	0·5	—	—	—
B vitamins					
thiamin (milligrams)	0·08	0·03	0·09	0·03	0·02
riboflavin (milligrams)	0·3	0·08	0·43	0·1	0·09
nicotinic acid (milligram equivalents)	1·8	1	2·5	0·5	0·5
pyridoxine (milligrams)	0·08	0·04	0·1	0·02	0·01
pantothenic acid (milligrams)	0·7	0·5	0·9	0·2	0·2
folic acid (micrograms)	7	5	—	—	—
biotin (micrograms)	4	0·4	4	0·8	—
vitamin B_{12} (micrograms)	0·8	0·2	0·5	0·1	Trace

a – value for unfortified milk. Fortified milk contains 0·7 micrograms
b – varies depending on mother's intake
— – no figures available
(1) taken from Documenta Geigy Scientific Tables 1970, Ciba-Geigy Ltd

ferable where possible. Diarrhoea is more likely when babies
are artificially fed: the feed is easily contaminated whilst it is
being prepared, and there is a protein in breast milk which
discourages the growth of harmful bacteria in the infant's
digestive system. It is possible that allergy to cows' milk
proteins is responsible for cot death—a rare tragedy more
common in artificial than breast fed babies. Convulsions,
developing in the first few days of life, are also more likely—
probably as a result of the high content of phosphorus in
cows' milk which renders most of the calcium and **magnesium**
unavailable for absorption into the infant blood stream.

In infancy the kidneys are less efficient than in adulthood,
requiring more water to eliminate excess **sodium, potassium**
and urea (the waste product of unwanted protein in the diet).
The breast-fed infant is able to eliminate waste products
easily (human milk has a low content of sodium and potassium, and its protein is of better quality (more efficiently
used) than cows). This may not be the case if the child is
artificially fed: more milk powder than is recommended by
manufacturers is frequently fed to babies. If the feed is concentrated and no extra water given the baby may become
dehydrated and cry, when the mother may aggravate the
accumulation of waste products by giving more feed. Convulsions and brain damage may follow. Diarrhoea, which
causes severe loss of water from the body, exacerbates
dehydration.

Obesity in both mother and child is more likely with artificial feeding. Appetite is probably stimulated by hormones
in pregnancy and most women accumulate about 9 lbs of
fat: an energy reserve probably specifically intended for
breast feeding. A nursing mother needs an average of 800
kilocalories of extra energy each day to produce her milk,
and only about 500 kilocalories can easily be eaten as extra
food. The additional 300 kilocalories is supplied from her fat
store. If she decides against breast feeding and does not
make a positive effort to lose weight by dieting, the extra
fat is not lost. Many overweight middle-aged women

attribute the beginning of their problem to their first pregnancy.

An artificially fed child given the correct amount of feed is not likely to become obese, but too many, or over-full scoops of milk powder (apart from the associated risk of dehydration) can result in a high energy diet. Too early weaning (sooner than 4 months) is also likely to cause obesity.

MILLET

Millet contains the same **nutrients** as other **cereals**. It cannot support life when eaten alone because it is lacking in vitamins C, A and B_{12}, but is a valuable staple in the diet.

AVERAGE NUTRIENTS IN 25 GRAMS (1 OZ) MILLETS

Nutrient	Bulrush millet	Finger millet	Sorghum
Energy (kilojoules)	380	350	370
(kilocalories)	90	85	90
Protein (grams)	3	$1\frac{1}{2}$	$2\frac{1}{2}$
Carbohydrate (grams)	15	20	20
Fat (grams)	$1\frac{1}{4}$	$\frac{1}{2}$	1
Calcium (milligrams)	5	90	10
Iron (milligrams)	0·75	1·25	1·13
B vitamins			
thiamin (milligrams)	0·08	0·08	0·13
riboflavin (milligrams)	0·04	0·03	0·03
nicotinic acid (milligram equivalents)	0·8	0·5	0·5
Vitamins A, C, B_{12}	0	0	0
Other nutrients	—	—	—

— – no figures available

There are several varieties of millets, each with different local names. The three main varieties are pearl or bulrush millet; red or finger millet; and sorghum. Finger millet contains over 10 times the **calcium** of wheat and about half the **protein**. Other millets have similar values to wheat, but all are lacking in **gluten**. Millet is eaten whole as a kind of thick gruel, or ground by hand to flour for use in flat cakes. As it is not highly milled, millet retains all its nutritional value. No commercial millet foods are produced.

Millet has been grown in most arid areas and used as a staple in communities in Africa, Asia and South America. Its cultivation is declining in favour of new drought resistant varieties of wheat which can be made into leavened bread.

MINERAL HYDROCARBONS

Products derived from mineral oil. They are undesirable in food: they interfere with the absorption (see Digestion) of the fat soluble vitamins A, D, E and K (large amounts of liquid paraffin can cause deficiency of these **nutrients**) and small amounts may be absorbed and deposited in the liver. Impurities are suspected carcinogens, consequently only high-grade products are permitted in food.

The Mineral Hydrocarbons in Food Regulations 1966 prohibit their use in all except six foods, but allow for contamination of all processed food (to a maximum of 0·2%) with mineral oil used as release agents (see Additives) to processing apparatus. The six foods that are otherwise permitted to contain mineral hydrocarbons are:

Dried fruit: mineral oil can be added to a maximum of 0·5% (500 milligrams per 100 grams of food) to keep the fruit moist. The regulations assume that dried fruit is to be washed before eating.

Citrus fruit: mineral wax can be added to the skin to a maximum of 0·1%. The regulations assume that the skin is not eaten.

Sweets: used as a polish or glazing agent (see Additives). Up to 0·2% is allowed.

Chewing gum: A maximum of 60% of mineral wax is permitted in the gum. Only very pure gum, subjected to a lengthy purification process, is allowed.

Eggs (must be marked 'sealed'): Mineral wax can be used to coat hen or duck egg shells, as a preservative—see Eggs.

Cheese: Mineral wax may be used to coat hard, Edam and Gouda cheese. It is usually removed before sale from hard cheeses, and painted red in Edam.

MINERAL WATER

Artificial—Water charged with the gas carbon dioxide. The effervescence is said to make the drink more refreshing than still water. Soda water must contain a minimum of 330 milligrams of sodium bicarbonate per pint.

Natural—Bottled spa waters. They have no proven benefits except that they may be drunk in preference to alcohol. Spa waters are usually alkaline and often have a strong smell or taste due to their mineral content. Some are charged with carbon dioxide.

MINERALS

Essential **nutrients** which must be supplied in the diet. They may be incorporated into the body structure, or function as cofactors for **enzymes**, or are required as **electrolytes**.

Humans require at least 18 minerals. Eight—**calcium, iron, magnesium, chlorine, phosphorus, potassium, sodium** and **sulphur** are found in relatively large quantities in food and the body. For example, adults contain about 1200 grams (over 2 lb) of calcium, and the adult daily **recommended intake** is 500 milligrams (one milligram is equal to one-thousandth of a gram). Comparatively minute amounts of the other minerals—the **trace elements**—are required and found in the body. For example adults contain about 300 milligrams of **iodine** and probably need about 100 micrograms a day (one microgram is equal to one-thousandth of a milligram).

MOLYBDENUM

An essential **nutrient** for animals and probably humans. It is a **trace element** and part of several **enzymes**, one of which is concerned in the formation of uric acid (see Gout) and possible in the utilisation of **iron**.

There are no well established values for molybdenum in food: the element is difficult to analyse and soil water and plant contents vary in different parts of the world. Leafy vegetables and pulses supply more than root vegetables. Plants grown in alkaline soils rich in humus absorb more than those grown in acid and sandy soils.

All diets supply some molybdenum. It is possible that the **teeth** of children living in a low molybdenum area may be less resistant to decay, but otherwise no disorder arising from lack of molybdenum is currently suspected.

In excess, molybdenum may possibly predispose to gout. Animals given feeds high in molybdenum become deficient in **copper**. A similar antagonistic effect of molybdenum is claimed to account for severe knock-knees of humans living in a certain area of India, where the diet is very restricted and the soil is alkaline and high in molybdenum and **fluorine**.

MONOSACCHARIDES

Simple sugars synthesised by plants, animals and microbes. There are many different monosaccharides in nature, but **fructose, glucose** and **galactose** are the most important in human nutrition. They are found only in small amounts naturally, but in abundance as combined forms—**disaccharides** and **polysaccharides**. During **digestion** the component monosaccharides of some combined forms in food are liberated and absorbed into the blood stream.

MONOSODIUM GLUTAMATE (MSG)

A **flavour** enhancer, which, like **salt**, intensifies the flavour of food. It also imparts a meaty taste. It is the sodium salt (in the same way that sodium chloride, common salt, is the salt of hydrochloric acid) of **glutamic acid**, an **amino acid** present in all food **proteins**. For instance, 100 grams (3 slices) of bread contain about 3 grams of glutamic acid, a medium sized (100 grams) grilled beefsteak about 4 grams. During **digestion**, proteins are dismantled and the amino acid slowly absorbed into the blood stream.

MSG is manufactured from wheat **gluten** and is an accepted part of Chinese cuisine. It has also been added to Western processed foods for about 70 years. Used in small amounts (100 to 200 milligrams per 100 grams of food) processed foods containing MSG are harmless for adults. However, MSG contributes to the sodium content of the diet and is therefore undesirable for babies (see Sodium).

MSG is absorbed more quickly than glutamic acid into the blood stream. Like other condiments and spices, excessive amounts may cause unpleasant symptoms in some people. Sweating, headache, numbness in the back of the neck, and chest pain have been reported by a number of people after eating Chinese food. Between 2 and 12 grams of MSG (10 to 120 times that present in processed foods) are necessary before the unpleasant effects are felt. It is absorbed more quickly (and therefore has a greater effect) from clear soups eaten on an empty stomach. Foods heavily laced with MSG may be served in a Chinese restaurant and the unsuspecting consumer may add soy sauce, which also contains MSG. Although the experience can be frightening for those who are sensitive to MSG, there is no evidence of permanent damage.

NICKEL

An essential **trace element** for some animals, and probably humans. In animals, it apparently helps to maintain the integrity of the liver. It is most unlikely that human diets would ever lack nickel: it is used as a catalyst for hydrogenating oils (see Margarine) and for coating steel food processing equipment. Some is likely to be transferred to food.

NICOTINIC ACID—other names, niacin, PP factor
group name for nicotinamide (also
called niacinamide) plus nicotinic
acid

Part of the **vitamin B complex**. Insufficient nicotinic acid or
tryptophan in the diet causes the deficiency disease **pellagra**.

The **amino acid, tryptophan**, which occurs in most **proteins**
can be converted to nicotinic acid in the liver when sufficient
tryptophan and vitamins B_1, B_2, and B_6 (**thiamin, riboflavin,
pyridoxine**) are available. Consequently food contents and
dietary requirements for the vitamin are expressed as nicotinic acid equivalents. One milligram equivalent is equal to
60 milligrams of tryptophan in food, or 1 milligram of
available nicotinic acid.

Most foods, except sugar, fats and alcoholic spirits, contain some nicotinic acid equivalents, but foods rich in protein
are the best sources. Liver, peanuts and **extracts** are rich
sources. Meats are good sources, followed by fish, cheese,
pulses, eggs, other vegetables, bread and milk. Fruit is a
poor source. During the roasting of coffee, nicotinic acid is
released from trigonelline, and both instant and ground
coffee are sources of the vitamin.

Cereals, except maize, contain tryptophan. Wholegrain
cereals also contain nicotinic acid (it is removed when cereals
are refined) but it is in a bound form, niacytin, not available
to man unless treated with alkalies. Bicarbonate of soda
(used to raise soda bread) may liberate some of the nicotinic
acid in wholemeal and brown soda bread, but it also destroys
thiamin (vitamin B_1). Ordinary (leavened with yeast) white
bread contains more nicotinic acid equivalents than wholemeal because nicotinic acid is added to all white flour by
law—see Bread. Some breakfast cereals also contain added
nicotinic acid.

In the average British diet (excluding meals taken outside
the home) in 1973 meat supplied most of the nicotinic acid
equivalents (32%) milk, cheese, eggs and fish 24%, cereals

20%, and vegetables 15%. Fruits supplied little, sugar and fats none.

Nicotinic acid is one of the most stable of vitamins and there are little or no losses during storage, processing and cooking. Weight-for-weight, cooked **meat** may contain more than raw. However, the vitamin is water soluble and leaches out into cooking water. Vegetables lose about 40% of the vitamin when boiled in a large pan of water, but none when baked or fried. The vitamin can be recovered if the cooking water is used for sauce or gravy.

The **recommended intake** for women is 15 milligrams of nicotinic acid equivalents per day: the average British diet supplies nearly twice this amount. Men need slightly more than women and, to ensure an adequate supply for the growing child, a mother probably needs more than usual throughout pregnancy and breast feeding. Infants, children and adolescents need more in proportion to their weight than adults to allow for growth. See Appendix for recommended intakes. The table facing shows portions of foods supplying the daily recommended intake of nicotinic acid equivalents for men (18 milligrams).

Nicotinic acid is the group name for two forms of the vitamin—nicotinic acid (also called niacin) and nicotinamide (also called niacinamide). Although nicotine in tobacco has a similar structure to nicotinic acid, the biochemical effects of the two are completely different. To avoid confusion, niacin is the preferred name for the vitamin in the USA. The vitamin is part of the B complex, but is not numbered in Britain for historical reasons (see Vitamin B).

The two forms of the vitamin occur in food and are equally potent. In the body they are converted to two important coEnzymes involved in the liberation of energy within cells. Although there is no store of nicotinic acid in the body, it is widely distributed throughout all cells. A diet totally lacking in both tryptophan and nicotinic acid takes about 50 days to deplete the tissues. The symptoms of pellagra—a sore tongue and diarrhoea; sunburn-like lesions on the skin; mental confusion and depression—are due to derangements of the digestive system, skin and brain cells. Death is usually hastened by an infection. Improvement of the condition results when the vitamin is given, but there are usually other deficiencies that must be treated before full health is restored.

AVERAGE PORTIONS OF SOME FOODS SUPPLYING THE DAILY RECOMMENDED INTAKE OF NICOTINIC ACID FOR MEN

	Foods	Weight containing an average of 18 milligram equivalents
Rich sources	Beef extract	35 grams (3 cups)
	Fried liver, peanuts	80 grams (3 oz)
Good sources	Kidney, cooked meats, salmon	180 grams (6¼ oz)
	Sardines, mackerel	215 grams (7½ oz)
	Other fish	280 grams (10 oz)
Moderate sources	Coffee, ground	180 grams (10 cups)
	Coffee, instant	40 grams (20 cups)
	Most cheese, almonds, mushrooms, walnuts	350 to 450 grams (¾ to 1 lb)
	Cottage cheese	560 grams (1¼ lb)
	Eggs	600 grams (10 standard)
	White bread	690 grams (17 large slices)
	Figs, dates, peas and other pulses	780 grams (1¾ lb)
	Fried potatoes, watercress	900 grams (2 lb)
	Milk	2 litres (3½ pints)
	Beer	2½ litres (4½ pints)
	Wholemeal bread, most biscuits, cakes, crispbreads, boiled potatoes	1 to 1½ kilograms (2¼ to 3½ lbs)
Poor sources	Most boiled vegetables, most fruits, boiled rice, cornflour, honey, double cream, wine	More than 2 kilograms (4½ lbs)

Fats, sugar, jams and alcoholic spirits contain no nicotinic acid

Pellagra is extremely rare in Britain but sometimes occurs when there is underlying disease, for example **malabsorption** and alcoholism. Nicotinamide is harmless when taken in excess—it is filtered out of the blood stream by the kidneys—but supplements are unnecessary in **balanced diets**. Nicotinic acid however sometimes causes transient flushing and a burning sensation in the head, neck and arms. Its use—in powders dusted over meat to preserve its red colour—caused a public controversy in 1963. It was subsequently banned from meat (together with vitamin C) by the Meat (Treatment) Regulations 1964.

In pharmacological doses (much more than could be obtained from food) nicotinic acid (but not nicotinamide) may suppress the synthesis of **cholesterol** and daily doses of 3 grams have been used in the treatment of some inherited disorders (see Lipid). Pharmacological doses of the vitamin are also used in two **Inborn Errors of Metabolism** and may be of benefit in the treatment of schizophrenia.

NITRITE AND NITRATE

Preservatives permitted in cheese (excluding cheddar, cheshire and soft cheeses), cured meat (including bacon, ham, luncheon meat) and pickled meat (like tongue, salt beef, corned beef, beef loaf, meat rolls, some sausages and paté). They are the sodium or potassium salts of nitrous and nitric acids respectively (in the same way that sodium chloride (salt) is the salt of hydrochloric acid).

In cheese, nitrite prevents souring. In cured and pickled meat it prevents multiplication of bacteria responsible for botulism (a type of food poisoning that is usually fatal); contributes to the flavour; and combines with the pigment (myoglobin) in meat forming a stable pink colour. Unlike canned meats, bacon and ham cannot be heated to a sufficiently high temperature to kill the bacteria without spoiling the product. **Ascorbic acid** (vitamin C) is used in modern curing methods and sometimes enables less nitrite to be added. Nitrate (in salt petre) was the original curing substance and is converted to nitrite in the meat, partly by bacterial action. Since it was realised that nitrite is the active substance, it is usually added directly. The Preservatives in Food Regulations 1974 permit a maximum of 10 parts per million of nitrite in cheese and 200 parts per million in cured and pickled meats (one part per million is equal to 0·1 milligrams per 100 grams of food).

Nitrates occur naturally in soil, water and some vegetables (they are the raw material of the protein made by plants). Abnormally high concentrations of nitrates in food and water can follow the excessive use of fertilisers. In itself this is harmless: nitrate is produced in the body and is toxic only in very large doses. However, enzymes from bacteria in vegetables and the human digestive system can convert the nitrate to nitrite.

Nitrite formation is more likely when little acid is secreted in the stomach, for example in infancy, and it combines with haemoglobin—the oxygen-carrying pigment in red blood cells. Unlike adults, babies are not equipped with enzymes capable of reversing the effect of nitrite poisoning (methaemoglobinaemia) and the condition can be fatal. Special water is available in high nitrate areas for babies; nitrite and nitrate are not added to baby foods.

In 1965, it was realised that nitrite in food might have other more harmful effects. Under certain circumstances, it reacts with amines (in protein containing foods but especially cheese and fish) to form nitrosamines, which are known to cause cancer in animals. Over 100 different nitrosamines are known (they are formed from a variety of amines) and they cause cancer in many areas of animals' bodies, including liver, lungs, brain and digestive tract. One can cause cancer in offspring when given to pregnant animals.

Although nitrosamines have been detected in human food, it should be stressed that the amounts found are at least 1000 times less than that needed to cause cancer in animals. Such small quantities may not cause cancer in humans (or, indeed, in animals) and in cured meats the additive is a necessary precaution against the fatal botulism. Nitrate is also found in saliva and nitrite (and therefore nitrosamines) could be formed even if not added to food. Food and vitamin C may inhibit their formation.

However, the situation is sufficiently serious to stimulate large-scale research programmes in many countries. There is too little evidence to support the banning of nitrite from food, but eating only moderate quantities of cured and pickled meats is probably a worthwhile precaution.

NITROGEN

Part of the structure of **amino acids**. The nitrogen content of a food is a measure of its **protein** content: proteins contain an average of 16% nitrogen. **Carbohydrates, fats** and **alcohol** contain no nitrogen.

Nitrogen is widespread in air (80% of the air is nitrogen gas): in combined forms in soil (including nitrates, ammonium salts) and in living matter (mainly protein). In excreta, nitrogen is mostly in the form of urea, the breakdown product of unwanted amino acids. Healthy adults are said to be in 'nitrogen balance': nitrogen eaten in food (mostly protein) is equal to nitrogen losses (mostly urea) from the body.

NOVEL PROTEIN

Artificial meat made from flavoured and coloured soya or—less commonly—field bean protein. There are other possible sources—see opposite.

Two types of novel protein are manufactured. The textured (extruded) type—usually called textured vegetable protein TVP—is basically a dried foam, made from soya flour, relatively cheap to produce and mostly sold to caterers and food manufacturers as chunks or granules for addition to meat products (like pies, stews, curries, sausages). Some is sold as a mince extender. At present novel protein is not allowed to replace the legal minimum meat content in **meat** products, but can be used in addition if declared on the label.

Spun protein is more expensive. The protein is extracted, treated with alkalies, and spun through small holes into an acid solution. The fibres produced are stretched, combined with binders and shaped. Convincing steak, chicken and ham are made. Different layers can be made into bacon. The texture of spun protein is superior to extruded, but as yet the

APPROXIMATE NUTRIENTS IN 60 GRAMS (2 OZ) NOVEL PROTEINS
(REHYDRATED)

Nutrient	Soya		Field bean	
	Extruded	Spun	Spun, Flavoured	Spun Unflavoured
Energy (kilojoules)	315	300	515	540
(kilocalories)	75	70	120	130
Protein (grams) (1)	11	14	10	10
Carbohydrate (grams)	7	4	1	0·1
Fat (grams)	0·1	0·1	9	10
Iron (milligrams) (2)	2·5	2·4	2·4	2·2
Potassium (milligrams)	500	285	2	1
Phosphorus (milligrams)	135	120	70	5
Magnesium (milligrams)	60	40	1	0·5
Calcium (milligrams)	55	40	5	5
Sodium (milligrams)	5a	480	65	2
Other minerals	Small amounts			
Vitamins A, C, D	0	0	0	0
B vitamins				
thiamin (milligrams) (3)	0·2–16	0·07	Trace–1	Trace
riboflavin (milligrams) (4)	0·07–0·5	0·1	Trace–0·04	0·01
nicotinic acid (milligram equivalents)	3–9	5	2	2
pyridoxine (milligrams)	0·16–0·78	0·2	0·04	0·04
pantothenic acid (milligrams)	0·4–1·4	0·3	0·03–0·06	0·06
folic acid (micrograms)	4–7	4	0·5	1
biotin (micrograms)	7–9	5	2	2
vitamin B_{12} (micrograms) (5)	Trace–3	1	Trace–0·5	0·5
Vitamin E	—	—	—	—

— – no figures available
a – salt added to flavoured products
(1) – recommended minimum level, 10 grams
(2) – recommended minimum level, 2 milligrams
(3) – recommended minimum level, 0·4 milligrams
(4) – recommended minimum level, 0·16 milligrams
(5) – recommended minimum level, 1 microgram

flavour of both types only resembles meat. If it is made clear
that the product contains no meat, foods made entirely from
novel protein can be sold.

When plants are processed directly into protein, by-
passing the animal link, more food is produced for the same
area of land. Novel proteins are thus cheaper than meat and
may eventually play the same role as margarine in diets.
Most contain as much or more protein, **minerals** and B
vitamins (except zinc, and vitamins B_1, B_2 and B_{12}) as meat.
Soya products contain little fat, though it is usually added to

spun products, and about a third of the dry weight is **carbohydrate**. However novel protein is of poorer quality than meat (it is less efficiently used for growth and repair when eaten alone because it is relatively short of the essential amino acid **methionine**) and its iron is less readily available to the body. The Food Standards Committee has recommended that the quality of novel proteins be improved by the addition of methionine and vitamins B_1, B_2 and B_{12}. It has also recommended minimum levels of protein, and iron for novel proteins.

Protein could be extracted and made into novel protein from other sources, like the residue left after oil has been extracted from cotton seeds, sunflower seeds, peanuts and coconuts; wastes from meat and fish production; green leaves; and micro-organisms (fungi, bacteria, yeasts and algae). Different strains of these micro-organisms are being developed to grow on a variety of wastes from the oil, gas, paper, and food industries. If eventually accepted as safe, microbial sources are initially intended for animal feeding— perhaps the most palatable way of increasing supplies of meat by technological means.

NUTRIENTS

Substances, contained in foods, which provide **energy** and raw materials for the synthesis and maintenance of living matter. Human nutrients are **protein, carbohydrate, fat, minerals** and **vitamins. Alcohol** provides energy but is a drug rather than a nutrient. Those that cannot be made in sufficient quantities from raw materials in food in the body are called essential nutrients: vitamins, minerals, essential **amino acids** and **linoleic acid**. Lack of an essential nutrient results in a specific deficiency disease. Excess quantities of some nutrients may cause ill health or reduce life span. Mixtures of foods that are thought to supply the optimum quantities of nutrients—so maintaining health—are called **balanced diets**.

Water and oxygen are also essential for human life, but are not usually classed as nutrients.

NUTS

Nuts are rich sources of **fat, protein, fibre** and some **minerals**. They cannot support life when eaten alone because they are lacking in vitamin B_{12} and usually vitamins C and A. Peanuts (botanically they are pulses) pine kernels, and melon, sunflower and sesame seeds are similar in nutritive value. Chestnuts differ from other nuts: they are low in protein and fat but high in **carbohydrate**.

Nuts contain more fat than fatty meat. Hazelnuts and coconuts contain about a third of their weight as fat: walnuts, almonds and peanuts about half; and brazils and desiccated coconut nearly two-thirds. Walnuts and sesame and sunflower seeds contain the highest percentage of polyunsaturated **fatty acids** and coconuts and cashews the lowest. Together with peanut butter (which is usually hydrogenated —see Margarine) coconut and cashews should be avoided in **cholesterol** lowering diets.

Most nuts are rich sources of **energy**, though chestnuts contain rather less than bread. 120 grams (4 oz) of peanuts supplies a quarter of the day's **recommended intake** for energy for most men. Nuts are avoided in most weight reducing diets, unless exchanged for meat or cheese.

Peanuts are richest in protein. Weight-for-weight they contain more than cheddar cheese; almonds, brazils, and walnuts more than eggs. However, when eaten alone, nut protein is used with less efficiency (i.e. is of poorer quality) than the protein in animal foods. Nuts are relatively lacking in the essential **amino acid lysine** and peanuts and almonds are also relatively lacking in **methionine**. These shortcomings are not important in normal mixed diets: the day's needs for essential amino acids are satisfied either by eating more nut protein, or by supplementing nuts with small quantities of animal protein (like cheese).

Average nutrients in 25 grams (1 oz) of some shelled nuts and seeds

Nutrient	Almonds	Brazils	Cashews	Chestnuts	Cobnuts	Coconut	Melon seeds	Peanuts	Pine kernels	Pistachio nuts	Pumpkin seeds	Sesame seeds	Sunflower seeds	Walnuts
Energy (kilojoules)	630	675	620	180	420	380	610	635	665	655	640	620	550	575
(kilocalories)	150	160	145	45	100	90	145	150	160	155	150	150	130	135
Protein (grams)	5	3½	5	½	2¼	1	6¼	7	3¼	5	7½	5	6¾	3
Fat (grams)	13½	15½	11¼	¾	9	9	11¼	12¼	15	13½	12½	12½	9	12¾
Cholesterol	0	0	0	0	0	0	0	0	0	0	0	0	0	0
Saturated fatty acids (as % of total) (1)	—	15	5	—	2	75	—	12	5	12	—	5	5	5
Polyunsaturated fatty acids (as % of total) (2)	20	25	7	—	10	2	—	30	60	20	—	40	65	75
Carbohydrate (grams)	1	1	—	9	1½	1	—	2	—	—	—	—	—	1¼
Potassium (milligrams)	215	190	—	125	85	110	—	170	240	—	—	—	—	170
Sodium (milligrams)	1	1	—	3	1	4	—	1	—	—	—	—	—	1
Phosphorus (milligrams)	110	150	—	20	55	25	—	90	125	—	—	—	—	125
Phosphorus as phytic acid (as % of total) (3)	82	86	—	18	74	81	—	57	—	—	—	—	—	42
Iron (milligrams)	1	0·7	1·3	0·2	0·3	0·5	4	0·5	1·3	3·5	2·5	2·5	1·8	0·6
Calcium (milligrams)	60	45	15	10	10	5	15	15	4	35	10	375a	25	15
Magnesium (milligrams)	65	100	—	10	15	15	—	45	—	—	—	—	—	35
Other minerals	Small amounts													
Vitamin A (microgram equivalents)	0	0	Trace	0	0	0	Trace	0	2	8	2	2	Trace	0
Vitamin D	0	0	0	0	0	0	0	0	0	0	0	0	0	0
Vitamin E (milligrams α tocopherol)	—	—	—	—	—	—	—	2	—	—	—	—	—	—
B vitamins														
thiamin (milligrams)	0·07	0·25	0·2	0·05	0·1	0·01	0·03	0·2	0·32	0·2	0·05	0·03	0·5	0·07
riboflavin (milligrams)	0·15	0·03	0·05	—	—	0·01	0·04	0·03	0·06	0·05	0·05	0·06	0·05	0·03

AVERAGE NUTRIENTS IN 25 GRAMS (1 OZ) OF SOME SHELLED NUTS AND SEEDS

Nutrient	Almonds	Brazils	Cashews	Chestnuts	Cobnuts	Coconut	Melon seeds	Peanuts	Pine kernels	Pistachio nuts	Pumpkin seeds	Sesame seeds	Sunflower seeds	Walnuts
nicotinic acid (milligram equivalents)	1·2	3	2	—	—	0·25	2·3	5·2	—	1·3	—	2·4	2·3	0·8
pyridoxine (milligrams)	0·02	0·25	—	—	—	0·02	—	—	—	—	—	—	—	0·25
pantothenic acid (milligrams)	0·02	—	—	0·07	—	—	—	0·67	—	—	—	—	—	0·17
biotin (micrograms)	0·1	—	—	0·3	—	—	—	7	—	—	—	—	—	0·5
folic acid (micrograms)	—	0	0	0	0]	0	0	0	0	0	0	0	0	12
vitamin B_{12} (micrograms)	Trace	Trace	0	Trace	Trace	Trace	Trace	0	0	0	0	0	0	0
Vitamin C (milligrams)	Trace	Trace	0	Trace	Trace	Trace	Trace	Trace	—	0	—	Trace	—	Trace

(1) – lauric, myristic and palmitic acids
(2) – linoleic acid
(3) – phytic acid interferes with the absorption of iron, calcium and magnesium
— – no analyses available
a – sesame seeds contain large amounts of *oxalic acid*

Nuts contain **calcium** (sesame seeds are richer than cheese) **magnesium** and **iron**. Some nuts are apparently better sources of these minerals than meat but sesame seeds contain **oxalic acid** and all nuts **phytic acid** which makes most of the calcium, iron and magnesium unavailable to the body. Nuts are rich in **potassium** which is well absorbed, and—unless salt is added during roasting—low in sodium.

Although nuts are deficient in vitamin B_{12}, they are at least as good a source as meat for others in the **vitamin B complex**. Brazils, peanuts, pine kernels, pistachios and sesame and sunflower seeds are as good sources of **thiamin** as pork, but about 75 % is lost on roasting. Peanuts are exceptionally rich in **nicotinic acid** equivalents, containing about as much as liver.

Mature nuts contain no vitamin C, but unripe (green) walnuts are richer than even rosehips, containing $1\frac{1}{2}$ to 3 grams of vitamin C per 100 grams. Almonds are richest in **vitamin E**: peanuts, brazils and sunflower seeds are also good sources. Pine kernels, pistachio nuts and sesame seeds contain a little carotene, but otherwise nuts are lacking in vitamin A.

The high fibre and fat content of nuts may account for their reputed indigestibility (fat increases the time food is left in the stomach) but when nuts are ground, or well chewed, their nutrients are fairly well digested.

Very few nuts (just over 1 ounce per month) are eaten in the average British diet. They are mostly used in confectionery and cakes. Peanuts are however an inexpensive source of nutrients, and all nuts can be a substitute for meat in vegetarian diets. Their lack of vitamin B_{12} is overcome by small quantities of egg, cheese or milk and more of their iron may be absorbed into the blood stream if citrus fruit, or other good source of vitamin C, is taken at the same meal.

OATS

A hardy **cereal** with a similar nutrient content to whole-wheat, except that it contains more fat and **biotin**. Only the husk is removed and oatmeal retains most of its **fibre** and all nutrients in the germ. Porridge oats are heat treated to inactivate a fat splitting **enzyme** in the germ which would otherwise free **fatty acids** that cause rancidity.

When made with 15 grams ($\frac{1}{2}$ oz) dried skim milk and 25 grams (1 oz) oats, a bowl of porridge contains more **protein, potassium** and vitamin B_2 **(riboflavin)**, and as much **energy**, vitamin B_1 **(thiamin)**, and **nicotinic acid** as two lean slices of grilled bacon. Oats contain **iron**, but also **phytic acid**: a glass of citrus fruit juice or other good source of

AVERAGE NUTRIENTS IN 25 GRAMS (1 OZ) OATS

Nutrients	in 25 grams dry oats	in a bowl of porridge made with 25 grams oats and 15 grams skim milk powder
Energy (kilojoules)	425	630
(kilocalories)	100	150
Protein (grams)	3	8
Fat (grams)	2	2
Carbohydrate (grams)	18	25
Potassium (milligrams)	90	290
Magnesium (milligrams)	30	45
Phosphorus (milligrams)	95	250
Calcium (milligrams)	15	205
Iron (milligrams)	1	1·08
Sodium (milligrams)	10	100
Other minerals	Small amounts	
Vitamin A	0	Trace
Vitamin C	0	1
Vitamin D	0	Trace
Vitamin E (α tocopherol, milligrams)	0·4	0·4
B vitamins		
thiamin (milligrams)	0·13	0·18
riboflavin (milligrams)	0·03	0·27
nicotinic acid (milligram equivalents)	0·7	2·2
pyridoxine (milligrams)	0·03	0·63
pantothenic acid (milligrams)	0·25	0·77
biotin (micrograms)	5	7·4
folic acid (micrograms)	3–15	—
vitamin B_{12}	0	0·3

— – no analyses available

vitamin C will increase the amount of iron absorbed into the blood stream.

OBESITY—medical name for the obese condition
—other name, overweight

Obesity is inevitable when the energy in food is regularly greater than energy needs. Only 50 kilocalories a day, from an extra knob of butter (or 10 minutes less time spent walking) can result in a fat gain of 40 kilograms (6½ stones) in 20 years. The superfluous energy is stored in congregations of fat cells—adipose tissue, or 'fat'—around organs and under the skin. Layers of fat under the skin distort the normal body contours. It is the most common nutritional disorder in affluent countries—up to half the British adult population is estimated to be obese.

Precise obesity tests measure the amount of fat in the body: normal women contain up to 30% of their weight as fat; men up to 20%. Most people however are judged obese if their weight is 10% or more above a desired weight (worked out by insurance companies) for their height and build. Using the table overleaf, a six-foot man of medium build weighing more than 88 kilograms (14 stones)—instead of his desirable weight of 79 kilograms (12½ stones)—would be obese. A critical look in a full-length mirror and the ruler rest (it should be possible to rest a ruler on both hip bones when lying flat) are alternative ways of confirming obesity.

Large fat deposits have been regarded as an asset in some societies, but are now unfashionable. More importantly, obesity reduces life span—a middle-aged man of more than 12½ kilograms (2 stones) overweight has a reduced life expectancy of 25%—and causes much disability. The excessive weight that has to be supported against gravity strains the legs and feet and may cause flat feet, arthritis in the knees (and knock-knees in children) and varicose veins. Fat around the lungs makes deep breathing difficult, contributing to bronchitis. Pendulous layers of fat are difficult to keep clean and dry, predisposing to infections, especially in the elderly. The heart enlarges to pump blood around the extra miles of blood vessels in adipose tissue and blood pressure may rise. Tiredness and lack of zest are due to fatigue—the extra weight that has to be carried around is often equivalent to a heavy suitcase. Overweight may precipitate **gout**, late onset **diabetes** and disorders of the blood **lipids**. These diseases are risk factors in **heart disease**.

Weight often accumulates gradually over the years, becoming obvious in middle life and after retirement, when

DESIRABLE WEIGHTS * FOR ADULTS OF MEDIUM[a] FRAME, AGED 25 AND OVER (IN INDOOR CLOTHING[b])

Height (no shoes)		Women		Men	
Metres	Ft/in.	Kilograms	St/lbs	Kilograms	St/lbs
1·42	4 8	44–49	6 12–7 9		
1·45	4 9	45–50	7 0–7 12		
1·47	4 10	46–51	7 3–8 1		
1·49	4 11	47–53	7 6–8 4		
1·52	5 0	49–54	7 9–8 7		
1·55	5 1	50–55	7 12–8 10	54–59	8 6–9 3
1·58	5 2	51–57	8 1–9 0	55–60	8 9–9 7
1·60	5 3	53–59	8 4–9 4	56–62	8 12–9 10
1·63	5 4	54–61	8 8–9 9	58–63	9 1–9 13
1·65	5 5	56–63	8 12–9 13	59–65	9 4–10 3
1·67	5 6	58–65	9 2–10 3	61–67	9 8–10 7
1·70	5 7	60–67	9 6–10 7	63–69	9 12–10 12
1·73	5 8	62–69	9 10–10 11	64–71	10 2–11 2
1·75	5 9	64–70	10 0–11 1	66–73	10 6–11 6
1·78	5 10	65–72	10 4–11 5	68–75	10 10–11 11
1·80	5 11			70–77	11 0–12 2
1·83	6 0			72–79	11 4–12 7
1·85	6 1			74–82	11 8–12 12
1·91	6 2			76–84	11 13–13 3
1·93	6 3			78–86	12 4–13 8

a – If of light frame, subtract 3 kgs from lowest and 5 kgs from highest desirable weights

If of heavy frame, add 4 kgs to lowest and 6 kgs to highest desirable weights

b – Women, allow 2 kgs for clothes, men 4 kgs.

* – Derived from Statis. Bull. Metropolitan Life Insurance Company **40** 1959

it may be dismissed as 'middle-age spread'. Men who were lean active sportsmen when young may become obese without gaining much weight. Their gain of fat will be masked by loss of muscle.

Despite the 20 tons of food eaten in an average lifetime, gradual accumulation of weight is neither a natural nor inevitable part of middle age. It is however more likely in those living a sedentary life. In the normal person there is a sensitive (but largely unknown) mechanism which strikes a balance between food eaten and energy used. When activity is reduced to a low level, the mechanism is thought to become insensitive and it becomes difficult to match the body's relatively low needs for energy with sufficiently little food. The difficulties are further increased by other roles of food: it is a necessary part of social life and can also be used as a palliative for boredom and unhappiness—sometimes as a result of habits formed in a childhood household particularly anxious about food.

An active job, or extra exercise (for instance at least an hour's brisk walk *every* day) are the best way of maintaining fitness in middle age. It may also be necessary to avoid large helpings of any food and eat little sugar and fat. Sugar, whether white or brown, should not be taken in drinks. Vegetables and fruits are low in energy but satisfying and can be taken in place of sweets, biscuits, cakes and heavy pastry or cream desserts. Bread and potatoes are not very high in energy and moderate quantities are preferable to large helpings of meat and some cheese. Fat or oil (when added as butter or in frying) doubles their energy content. Cocktail snacks (especially crisps and peanuts) should be eaten sparingly. A regular drinker would find it difficult to avoid obesity: 4 pints of beer (or half a bottle of spirits) supplies over a quarter of daily recommended intake for energy for a man living a sedentary life. These precautions are usually sufficient to avoid middle-age spread, but it is advisable to check weight every two months and start a slimming diet immediately if more than about 4 pounds are gained. Most people find it easy to lose a few pounds of weight but lack the will-power to lose more.

Many obese middle-aged women blame their first pregnancy for the beginning of their problem. There is a natural tendency to accumulate fat in pregnancy and it is usually not necessary to consciously 'eat for two'. If the child is not to be breast fed, a slimming diet should be started as soon as possible after the birth. Successive gains of weight with each child can result in tenacious fat in middle age.

Obesity in infancy, childhood and adolescence is particularly harmful. If overweight develops during growth the child has a greater chance of becoming a grossly obese adult, possibly because more fat cells are formed. Artificially fed babies are more likely to become plump: too many or overconcentrated feeds are easily given (see Milk). Too early weaning (before four months) can also cause obesity. The fat is likely to be carried on into childhood. It is better to slim the child before he becomes a fat adolescent (an unhappy adolescent who tries to slim may become anorexic —see Starvation), but fat children should not be made to feel guilty about their weight. Parents who confuse the relationship between hunger and eating are often the root cause of a fat child. Tactful attempts should be made to encourage the child to eat less, and if possible the parents should also diet, when no cakes, biscuits, pastry, soft drinks, puddings and fried food would be in the home to tempt the child. Medical and qualified dietetic advice should be sought if this fails: a strict (800 to 1000 kilocalorie) diet is usually necessary.

Grossly obese adults require an almost superhuman will-power to lose all their fat—they may have up to 75 kilograms (12 stones) of fat to lose—and rarely manage to become permanently slim. They may also have an added handicap: the burden of weight causes fatigue and reduces activity, lowering the daily needs for energy below average. Thus obese people are not necessarily gluttons (and often appear to eat no more than normal people). Nevertheless, at some time they have eaten more than their individual needs.

Obesity often runs in families and, although it is likely that attitudes to food learnt in the home are very important, there may be a genetic predisposition. Identical twins brought up by different sets of adoptive parents tend to reach adult weights closer to each other than their adoptive parents. The difference is not solely due to disparity in basal metabolic rate (see Metabolism)—some obese people have a high metabolic rate—nor to hormonal defects (which only rarely cause obesity). Neither do obese people absorb more energy from their food during digestion. It has been suggested that some people are able to 'burn off' excess energy in food increasing the amount of heat dissipated if their food contains more energy than required. When overfed, some people apparently find it hard to put on weight; others gain weight easily. The presence of an ability in thin people (or its absence in the obese) to maintain their weight by this mechanism—thermogenesis or luxus consumption—remains controversial. If it does exist, a drug which stimulates thermogenesis in the obese would make slimming diets unnecessary.

OILS

Vegetable oils rank with cooking fats, butter and margarine as the richest source of **energy** in the diet. They are 99·9% **fat**. Most are rich sources of **vitamin E** but the oils contain no other vitamins or minerals (except for red palm oil which contains carotene (pro-vitamin A)). Oils contain no **cholesterol.**

Most oils have a higher percentage of polyunsaturated **fatty acids** compared with animals fats and a low percentage of the saturated fatty acids known to raise the blood cholesterol. Safflower, sunflower, soya, corn and groundnut (arachis) oils are suitable for blood cholesterol lowering diets: they are low in saturated fatty acids and more than 30% of their fatty acids are polyunsaturated. Olive oil

APPROXIMATE NUTRIENTS IN 10 GRAMS (1 DESSERTSPOON) OF SOME VEGETABLE OILS

Nutrient	Coconut	Corn (maize)	Cotton-seed	Olive	Palm	Peanut (ground-nut)	Rape seed	Saf-flower	Soya	Sun-flower
Energy (kilojoules)	370	370	370	370	370	370	370	370	370	370
(kilocalories)	90	90	90	90	90	90	90	90	90	90
Fat (grams)	10	10	10	10	10	10	10	10	10	10
Cholesterol	0	0	0	0	0	0	0	0	0	0
Saturated[a] fatty acids (as % of total)	75	10	25	15	45	15	5	5	10	5
Stearic acid[b] (as % of total)	2	3	5	2	5	3	1	2	2	5
Monounsaturated[c] fatty acids (as % of total)	5	35	20	70	40	50	15	15	25	25
Polyunsaturated[d] fatty acids (as % of total)	2	50	50	10	10	30	15	70	55	65
Vitamin A (micrograms)	0	0	0	10	red 30–3000	0	0	0	0	0
Vitamin E[e]—α tocopherol (milligrams)	0·05	1·5	4	0·5	2·5	1	1·5	4	1	5
—other forms (milligrams)	0·2	5·5	2	0·5	7	2	4	1	8	0·5
Other vitamins and minerals	f	f	f	f	f	f	f	f	f	f

a – total lauric, myristic and palmitic fatty acids
b – Stearic acid is a saturated fatty acid, but has less effect on the blood cholesterol than lauric, myristic and palmitic
c – mainly oleic acid
d – mainly linoleic acid
e – oils vary greatly in their natural content of vitamin E, and there are variable losses during refining
f – 0, or traces only

(which is low in both saturated and polyunsaturated fatty acids) and cotton seed oil (which is moderately high in both) probably have a neutral effect. Cocoa butter, palm oil and coconut oil are high in saturated fatty acids. Large quantities of rape seed oil should be avoided: some contain a high percentage of erucic acid which caused heart damage in animals fed large amounts. Rape seed oil is usually included in 'vegetable cooking oil' blends.

Polyunsaturates in oils are more susceptible to rancidity than saturated fatty acids but vitamin E (a natural **anti-oxidant**) in fresh oils delays its onset. **Trace elements** (like copper) and free fatty acids which hasten rancidity are removed from most oils on sale. It is advisable to store oils in a dark cool place: light and warmth also promote rancidity.

There is little loss of nutritive value when oils are used in baking, but polyunsaturates decline when oils are heated for frying. Probably 10 to 20% are lost over a short period of frying (for example with chips), but the losses are greater with longer cooking times and higher temperatures. Overheating or repeated use of oil for frying causes free fatty acids to be released from **glycerol**. They lower the temperature at which fats smoke, when acrolein—responsible for the unpleasant flavour of burnt fat—is formed from glycerol. Drastically overheated fats have been shown to be toxic to laboratory animals. Although there is no evidence that rancid or cooking oils overheated in the home are a danger to health, they contain no vitamin E or polyunsaturated fatty acids.

Other oils

Essential oils are the flavours (or essences) extracted from plants, like oil of peppermint, oil of lemon, oil of almond—see Flavour. They contain many complex substances and most are not digested. Fusel oils (congeners) are responsible for the flavour of alcoholic drinks.

Mineral oils have a different chemical structure to other oils and fats and cannot be digested, passing out of the system with other undigested food. Vitamins A, D, E, and K dissolve in liquid paraffin (a laxative that lubricates the stools) and excessive use can cause deficiencies. Mineral oil is added to some food—see Mineral hydrocarbons.

Castor oil extracted from the castor oil bean is composed of simple fat. It is digested in the normal way, but ricinoleic acid (a fatty acid) is liberated irritating the muscles of the digestive system, causing purgation.

OSTEOMALACIA

The adult form of rickets: both diseases are usually caused by lack of **vitamin D**. Insufficient **calcium** is absorbed from food, and, to maintain a normal blood level, calcium is withdrawn from the bones. The skeleton is weakened, resulting in persistent pain, deformity, and poor healing of fractures. If not corrected, muscular weakness and tetany (twitching) occur when the blood calcium falls.

In Britain, osteomalacia can complicate **malabsorption** syndromes and some kidney disease, but it is primarily an important problem in the elderly and Asian immigrants. Confinement inside the house and heavy or traditional clothing when outside prevent sunlight from reaching the skin and initiating vitamin D synthesis. Once osteomalacia has developed, pain makes walking difficult and excursions from the house are even less frequent, completing a vicious circle. The elderly may require more vitamin D than young adults, perhaps due to the natural decline in the efficiency of the kidney (the kidney converts vitamin D to its active form).

If sunlight is not available, the needs for vitamin D are best met by using margarine, instead of butter or oil, and one multivitamin capsule a day. Excessive vitamin D is toxic.

OSTEOPOROSIS

Loss of bone. It is a very common disease of old age,
responsible for easily fractured bones and—due to compres-
sion of the backbone—loss of height. It is probably the most
common cause of backache in the elderly (though many
people experience no pain) and is often combined with
osteomalacia.

Everyone begins to lose bone in early middle age, perhaps
because the production of sex hormones begins to decline
at this time. Those who have grown more bone by adulthood
may lose the same percentage as others, but—because there
is sufficient left in old age—do not develop severe symptoms.
Women are at greater risk than men: they form less bone
and lose it at a faster rate.

Improvement of diet in adulthood probably does not
affect the course of the disease, but a childhood diet poor in
calcium, and perhaps **fluorine**, may result in a less dense
skeleton and greater likelihood of osteoporosis in old age.
Immobilisation (for example confinement to bed) and
probably a sedentary life increase the rate of bone loss.
Vegetarians may be less affected than meat eaters perhaps
because the acid residue of meat (see Ash) increases the rate
at which calcium is excreted by the kidney. Synthetic
oestrogens given after the change of life may slow the rate of
bone loss, but do not prevent it.

OXALIC ACID

A simple acid in foods which forms insoluble salts with
calcium and **magnesium** in the digestive system, rendering
these minerals unavailable to the body. Rhubarb leaves
contain toxic quantities.

Tea contains 15 milligrams per cup, fruit and cereals 0 to
20 milligrams, vegetables 0 to 50 milligrams, and meat and
fish less than 5 milligrams per 100 grams (3½oz). Beetroot,
rhubarb, spinach, cocoa and unripe tomatoes contain up to

500 milligrams per 100 grams. The average diet contains 70 to 150 milligrams of oxalic acid, but only about 10 to 20% of this is absorbed and later excreted in the urine, together with oxalic acid made from **glycine** and **vitamin C**. The rest is passed out in the stools as insoluble salts.

In some people excessive oxalic acid can be made in the body from glycine and be passed out of the body in the urine. Unless the urine is kept very dilute (by drinking a lot of water) insoluble calcium oxalate crystals form and eventually solidify into kidney stones. Very large doses of vitamin C, excessive tea drinking, and spinach and other high oxalate foods are probably best avoided.

PANTOTHENIC ACID—obsolete name, vitamin B_3

Part of the **vitamin B complex**. It is however found in all foods—except sugar, fats and spirits, and human deficiency has only been induced under volunteer experimental conditions, using a synthetic diet.

The richest sources of the vitamin are yeast, yeast extract and liver (which contains 8 milligrams per 100 grams ($3\frac{1}{2}$ oz)). Kidney, sweetbreads, heart, brains, egg yolk, peanuts, wheat germ and mushrooms are good sources, containing over 2 milligrams per 100 grams. The Royal jelly of bees contains 1 to 3 milligrams per 10 grams. Milk contains about 2 milligrams per pint, wholemeal bread about $1\frac{1}{2}$ milligrams per 100 grams. Refined cereals are poorer than wholegrain— about half is lost when wheat is milled to white flour. Other foods (cheese, fruit, other meat and vegetables, and fish) supply up to 1 milligram per 100 grams.

Pantothenic acid leaches out of foods into cooking water, like other B vitamins, and is also sensitive to dry heat, acids and alkalies. About a third is lost when meat and vegetables are cooked. Probably all the vitamin is retained during freezing, but a further third can be lost in the drip of thawed meats. Between one-third and three-quarters may be lost when foods are canned.

No **recommended intake** for the vitamin has been set, due to lack of information, but it is estimated that adults need between 5 and 10 milligrams per day, an amount contained in the average British daily diet.

Pantothenic acid is not stored, but it is present in all cells and a diet lacking in the vitamin takes about 3 months to deplete the body. As part of co**Enzyme A** (a vital link in the chain which liberates **energy** from food for synthesis of new substances needed for maintenance of life) pantothenic acid has a central role in **metabolism** and deficiency in animals and man results in a variety of symptoms. Human deficiency symptoms included headache; serious personality changes like irritability and restlessness; fatigue after very mild exercise; disordered sensation (pins and needles, etc.); stomach cramps; and difficulty in walking.

Despite its importance, there are no specific diseases attributed to lack of pantothenic acid. It has been used in the 'burning feet syndrome'—which develops in famine conditions and causes aching throbbing feet and stabbing pains in the legs and feet—but there are usually other dietary deficiencies which must be treated before full health is restored. Lack of the vitamin causes grey hair in rats and mink, but pantothenic acid is ineffective in preventing greying of human hair.

PARA AMINO BENZOIC ACID (PABA)

Part of the vitamin **folic acid**, which is synthesised from paba by micro-organisms. Paba is therefore an essential **nutrient** for micro-organisms but not humans, who depend on a dietary supply of folic acid. The sulphonamide drugs are toxic to some bacteria: they have a similar structure to paba and interfere with the formation of folic acid by bacteria, which cannot multiply in the absence of this vitamin. The curative effect of sulphonamides in diseases (like typhoid) caused by bacteria is negated if paba is also taken. Lack of paba causes grey fur in black rats, but not grey hair in humans.

PECTIN

A group of indigestible **polysaccharides** in fruits and vegetables which contributes to the dietary fibre in food. It is commercially extracted from apple pulp or citrus fruit pith and used as an **emulsifier** and stabiliser, and in the manufacture of jam (see Preserves).

The cells of fruits and vegetables are held together by a layer of pectin with calcium attached. It is degraded as fruits ripen and by cooking. The cells separate, softening the food. Calcium and magnesium salts, which retard the breakdown, are added (and called firming agents—see Miscellaneous **additives**) to fruits and vegetables like tomatoes and strawberries which would become very soft during canning and pickling. Some vegetables have to be canned in calcium-free water, otherwise the food would be tough. Alkali (like bicarbonate of soda) softens vegetables (by hastening the breakdown of pectin) when added to cooking water, but also destroys **vitamin C** and **thiamin** (vitamin B_1).

PELLAGRA

A deficiency disease caused by a diet lacking in **nicotinic acid** and the **amino acid tryptophan**. In Britain it occurs when there is underlying disease, like chronic alcoholism, and **malabsorption**.

Initial symptoms are loss of weight, depression, weakness and inability to concentrate. The skin is red and itchy, later scaly, brown and infected, particularly over the face, neck and hands, but also in any area of the skin exposed to hot fires or sun. In chronic cases there are severe lesions of the brain and nervous system, causing dementia and difficulty in walking. A sore mouth and tongue make eating difficult, and there may be diarrhoea or constipation. Only 24 hours after taking the vitamin, skin and digestive disorders are improved, though the nervous system may be irreversibly damaged. There are usually other deficiencies which must be remedied before full health is restored.

Nicotinic acid was known to chemists by 1840, but there was a 100-years' delay in identifying its connection with pellagra, due to several confusing factors. As late as 1940, at least 2000 people died from pellagra in the USA. The disease was endemic in many communities subsisting on **maize** but chemical analyses revealed that maize was an apparently good source of the vitamin. The disease could be prevented by **protein**, especially **milk** (which is not a particularly good source of the vitamin) and did not occur when the diet contained plenty of animal protein. However, it was less common in communities subsisting on other cereals (like wheat) where no more animal protein was eaten than in those where pellagra was rife; in coffee drinking communities; and in those where maize was cooked with lime. Solution of the problem came with the realisation that nicotinic acid is a vitamin; that tryptophan (lacking in maize but adequate in other cereals and well supplied in animal protein) was converted to nicotinic acid in the body; and that nicotinic acid is in a bound form (niacytin) in all cereals and cannot be digested unless released by treatment with alkali, like lime. Other cereals were protective because they contained tryptophan. Coffee is a source of the vitamin. Today, nicotinic acid equivalents take account of the contribution of tryptophan.

PHENYLALANINE

An essential **amino acid**—it cannot be made in the body and has to be obtained from **proteins** in food. It is used for the synthesis of new body protein needed for growth and repair and is also the precursor of some hormones and the pigment melanin in hair, eyes and tanned skin. **Tyrosine**, made from phenylalanine can partially replace phenylalanine in the diet.

It is abundant in food proteins and deficiency of phenylalanine alone never occurs when a normal diet is eaten: four large slices of bread for example supply the estimated adult daily needs for phenylalanine and tyrosine. Excesses are converted to **glucose** or **fat** and used for **energy**. The **nitrogen** part is converted to urea, later filtered out of the blood stream by the kidneys.

Phenylketonuria is the commonest **Inborn Error of Metabolism** successfully treated by diet. The absence of an **enzyme** in the liver blocks the normal metabolism of phenylalanine and the brain is irreversibly damaged unless a diet low in phenylalanine is given in the first few weeks of life. A synthetic diet, containing all nutrients except phenylalanine, is necessary together with low protein foods high in energy, like sugar, butter and low protein bread. The child's needs for phenylalanine for growth are given as small carefully calculated and weighed portions of normal foods, like milk. The diet is very specialised but the effort necessary to overcome the difficulties is well worth while. Treated phenylketonuric children are healthy and of normal intelligence. It is thought safe to discontinue the diet before adolescence (after the brain has matured) but it must be reinstated during pregnancy to avoid damaging the growing child.

PHOSPHOLIPIDS

Important substances concerned with the transport of fat (**lipid**) in the blood stream and probably the orderly organisation of molecules in the protoplasm of cells. Phospholipids form part of cell membranes and are especially abundant in nervous tissues, where they are incorporated into the myelin layer around fibres. This layer (or sheath) is an insulator—preventing leakage of nervous impulses from one nerve to another.

There are many different phospholipids, but their basic structure is a composition of **glycerol, phosphorus,** and up to two **fatty acids**. Some may have **choline** or serine (an amino acid) attached to the phosphorus part. **Lecithin** is the most common.

Phospholipids are abundant in both plant and animal foods—they are part of the structure of cells—but the body makes its own phospholipids and dietary supplies are unnecessary. The phospholipids (like glycerol phosphate) added to 'tonics' are of dubious value.

PHOSPHORUS

A **mineral**. It is a vital constituent of cells and, with **calcium**, needed for the hardening of bones and teeth.

All foods, with the exception of fats, sugar and spirits contain some phosphorus, but good sources of calcium and protein (like milk, cheese, meat, fish and eggs) are richest. Cereals, nuts and pulses are apparently good sources but most of it is in the form of **phytic acid** unavailable to the body. Phosphorus is also added by food manufacturers as food **additives** (for example as **polyphosphates**, **raising agents**, and phosphoric acid in some **soft drinks**).

There is no **recommended intake** set for phosphorus: diets which supply the recommended intakes of **protein** and calcium contain sufficient. Adults eat between one and two grams a day, of which about two-thirds is absorbed into the blood stream. An equivalent amount is later removed from the body in urine by the kidneys. However, infants, children and adolescents retain phosphorus for new cell and bone formation. It is also retained for the developing child in pregnancy and breast feeding.

When adulthood is reached, most people have accumulated more than 600 grams (over 1 lb) of phosphorus. Nearly all (80 to 85%) is in bone, combined with calcium, but it has other important roles, one of which is concerned with the transfer of energy (see Metabolism) inside cells. Energy released from **glucose** and fats (see Lipid) is transferred into a readily available form as 'high energy' phosphate bonds, most importantly in adenosine triphosphate (ATP). When

energy is required—for example if muscles are to contract—
enzymes break off the phosphorus and energy is liberated
from the bond.

Lack of phosphorus alone from an inadequate diet does
not occur in humans, although it is a problem in animals.
Temporary inadequacies in the diet are counteracted by the
kidneys, which cut down the quantity excreted in urine.
However, despite conservation by the kidneys, phosphorus
depletion has been caused by excessive quantities of some
antacids used for digestive and certain kidney disorders.
They bind phosphorus in food and none may be absorbed
into the blood stream. Phosphorus is withdrawn from the
bones, which become so weak and painful that walking
becomes very difficult. Lack of appetite, and severe tiredness
and weakness follow. Death is inevitable unless the antacids
are changed.

Excess phosphorus is probably harmless for adults—it
is either not absorbed or is eliminated by the kidneys.
Phosphorus containing 'tonics' are of dubious value:
weight-for-weight most foods are better sources of phosphorus. Many tonics rely on their alcohol content for any
apparent improvement in morale.

PHYTIC ACID

Inositol combined with **phosphorus**. During **digestion** it forms
insoluble salts with essential **minerals** (like calcium, iron,
magnesium and zinc) in food, rendering them unavailable
for absorption into the blood stream. Up to 80% of the
phosphorus in **nuts**, wholegrain cereals and **pulses** is in the
form of phytic acid, but nearly all is removed (with the bran)
from white flour.

At one time, phytic acid was implicated as a cause of
rickets but this is now thought unlikely. The body appears
to adapt to the phytic acid in brown and wholemeal bread
and other cereals, and eventually more calcium is absorbed
out of food. It seems that an **enzyme**, secreted in digestive

juices, is able to degrade phytic acid, splitting it to inositol and phosphorus. Phytic acid is also partially degraded by enzymes in yeast during proving of **bread**, and in pulses when they are soaked in water. Little is spilt during new bread making methods, which do not include a period of proving.

The extent to which phytic acid interferes with the absorption of other minerals—magnesium, iron and zinc—remains uncertain. A high phytic acid diet (from unleavened wholemeal breads) is thought to be a cause of zinc deficiency.

The extraction rate of **flour** was increased during wartime, and calcium (as chalk) was added to all flour except wholemeal by law to overcome the then suspected rickets-producing effect of brown bread. These regulations are still in force and white flour now contains over three times more calcium than that needed to overcome the effect of phytic acid. The regulations have not been revoked in case calcium has a protective effect in heart disease (see Water).

POISONS IN FOOD

A food may contain natural toxins, and is liable to contamination with unwanted substances at several points before actually reaching the plate. Pollutants in the environment can contaminate food before it is harvested; potentially toxic substances can be intentionally added on the farm or in transport and storage; and food may be made unwholesome during food processing or in the home. Most contaminants can get into food at more than one of these stages (for instance both on the farm and in the home) and problems of assessing the risk to consumers' health are similar for both natural and added contaminants. For these reasons, harmful substances in food have been grouped into this one section under the following headings:

Chemicals—Pesticides
          Radiations

Trace elements: arsenic, cadmium, mercury
and lead
Miscellaneous
Poisons in plants and moulds

Food poisoning Large doses of a contaminant can have an
immediate (acute) effect, like vomiting or fits, when it is easy
to pinpoint the offending substance and work out how much
is toxic. However, much smaller doses may be taken with no
apparent harm and it is difficult to assess whether at these
levels the substance is a poison. The body may be able
to cope with small regular intakes, either eliminating it
immediately or converting it to a harmless substance. Any
damage that might have been done can then be repaired.
The problems are greater when substances are eliminated
with difficulty: although they may be stored out of harm's
way (for instance in body fat) they do accumulate with
increasing age. The risks from carcinogens, mutagens and
teratogens (see below) are even more difficult to establish.
They react with DNA, the genetic blueprint in cells, causing
a change in the genetic material. Carcinogens induce
cancer: mutagens alter the germ cells (sperm and ova), when
the change is passed on to succeeding generations. The
mutation may be harmless, or at the other extreme be
incompatible with life. Teratogens damage the growing child
in the womb, and may cause severe mental or physical
abnormalities. Thalidomide is an example of a teratogen.
Other substances may contribute to other chronic diseases,
like heart disease.

Acceptable daily intakes (or 'safe' limits) below which
there is no measurable harmful effect can usually be
established for individual toxicants. However, they are
largely based on animal experiments which may not apply
to the human situation and it is only possible to say with
certainty that food intakes below the 'safe' limit *probably*
carry no risks to health.

Chemical contamination

Pesticides. Insecticides, herbicides and other substances added to food during farming, transport and storage before processing. There are about 100 in use, and some examples

TABLE 1. EXAMPLES OF SOME PESTICIDES

Type	Examples	
Insecticides	Derris, pyrethrum, nicotine	Extracted from plants
	Lead arsenate, calomel	Calomel contains mercury
	DDT, aldrin, dieldrin[a], BHC, endrin, chlordane	Organochlorine compounds
	Parathion, malathion, dichlorvos, TEPP	Organophosphorus compounds
	Tar oil	
Fungicides	Sulphur, copper sulphates	
	Organo mercury compounds	
	Dithiocarbamates	
Herbicides	Sodium chlorate	
	MCPA, MCPB, 2, 4, 5T	Hormonal
	Paraquat, diquat	
Rodenticide	Warfarin	

[a] – Converted to aldrin in the environment

are shown in Table 1. Most pesticides are acutely toxic in relatively large doses and, although they are assumed to be safe at present levels of intake, animal experiments suggest that some might be mutagens, carcinogens or teratogens at higher levels. Nevertheless, it is generally felt that the benefits of pesticides outweigh these risks: without their use at least a quarter of food would be lost.

The most frequently used insecticides are DDT, aldrin, and BHC, which are chlorinated hydrocarbons. DDT is not acutely toxic, in contrast to other chlorinated hydrocarbons which have sometimes poisoned humans. In 1956, 30 people had fits after eating bread made from flour that had been transported in a railway carriage, previously contaminated with endrin.

However, all the chlorinated hydrocarbons are particularly hazardous over the long term. They have a stable chemical structure, only slowly converted to harmless substances, and are continually recycled in the environment. Thus, although only used directly on vegetables and fruit (which should be washed or peeled before eating), all foods contain DDT. Animal fats supply the most. Chlorinated hydrocarbons are stored in the body fat and excessive accumulations stimulate certain liver enzymes, hastening the breakdown of some drugs and oestrogens. Animals highest in the food chain, like predatory birds and humans, are at greatest risk of accumulating too much. Since the early 1960s—when many birds were threatened with extinction—the use of DDT has been controlled and average intakes have been falling. British average intakes of DDT and BHC are now well below the 'safe' limit, but intakes of aldrin, may be slightly higher.

Other pesticides, except those containing lead, mercury and arsenic, see below, have fewer long-term problems, being detoxified fairly quickly. However, they can be carelessly handled. The organophosphorus compounds— derived from 'nerve gases'—are lethal and have poisoned several agricultural workers. In 1959, 9 children died in Singapore after eating barley that had been in contact with parathion.

Radiation

Radiations are given off by radioactive substances— radioisotopes—as they decay to more stable forms. The radiations are carcinogens, mutagens and teratogens. There are many natural radioisotopes—for instance one of potassium (K 40), another of carbon (C 14), and uranium (U 236)—in the environment, which together with cosmic rays, contribute to 'background radiation'. Technology has also resulted in other sources of radiation, for instance X-rays used in medicine and from TV sets, and radiation from artificial radioisotopes.

Artificial radioisotopes are formed in nuclear reactions,

for instance, in atomic weapons and nuclear power stations. Except when there is an accident, artificial radioisotopes from power-stations do not significantly contaminate the environment, but they are released when atomic weapons are exploded in the atmosphere. They eventually return to the ground, mostly in rain, as fall-out.

Both water and soil are contaminated with fall-out, but as plants absorb relatively little from soil, the outer parts—in direct contact with rain—are the most radioactive. Thus outer leaves are more radioactive than inner, and wholemeal flour is more radioactive than white. Food from animals fed on contaminated pastures also contributes to the radio-activity of diets: local diets in hilly areas with west facing slopes (which have the most rainfall) are more contaminated than the average. Most radioactivity is filtered out in soil and drinking water tends to be low. Fish also are relatively unpolluted with radioactivity because rain water is diluted by sea water.

The most important artificial radioisotopes likely to contaminate food are shown in Table 2. Strontium 90 is very hazardous over the long term because it has a long half-life and—being chemically similar—is used almost as if it were **calcium** by the body. It is consequently retained in bone, increasing the risk of cancer of the bone marrow, which is one of the most radiation-sensitive body tissues. Foods high in calcium contain the most strontium 90, but to some extent calcium is protective. Living things tend to use calcium in preference to strontium 90 so that plants grown in limed soil absorb less than others, and calcium added to white **bread** fortuitously reduces the amount of strontium 90 entering the blood stream after **digestion**.

Soft tissue, like the reproductive organs (which are also very sensitive to radiations) are likely to be damaged by caesium 137, but it is less dangerous than strontium 90. It is chemically similar to **potassium** and like potassium is eliminated fairly rapidly from the body. Foods high in potassium (like meat) are significant sources, but to some

TABLE 2. ARTIFICIAL RADIOISOTOPES IN FOOD

Radioisotope	Half life[a]	Chemically similar to	Most damaging to	Main food source in the average diet
Iodine 131	8 days	Iodine	Thyroid gland	Milk—cheese and butter do not contain I 131 because it decays during their manufacture
Strontium 90	28 years	Calcium	Bone marrow	Dairy products, cereals
Caesium 137	30 years	Potassium	Any soft tissue, but especially the germ cells	Dairy products, meat

[a] – the time taken for the radiations to decrease by half

extent potassium is also protective against caesium 137. Although wholemeal flour contains more than white, less is absorbed into the blood stream because wholemeal flour is higher in potassium.

Iodine 131, which like **iodine** is retained in the thyroid gland, is the most dangerous radioisotope after a nuclear power station accident or atomic explosion, but is not hazardous over the long term. Nearly all decays (to iodine) within a month. In the event of an accident it contaminates the immediate vicinity of a power station, when local milk supplies are temporarily withdrawn (as happened at Windscale in 1957). The government has stocks of iodine 131-free milk in readiness for emergencies.

The radioactivity of food, air, water and bone is closely monitored by various government bodies like the Agricultural Research Council, the Atomic Energy Authority, and the Medical Research Council. Fall-out reached a peak in 1963, following massive atomic weapons testing in the atmosphere, and declined rapidly until 1967. Since then there have been sporadic testings by some countries and the decline has levelled out. Compared with background radiation, radiations from artificial radioisotopes are now fortunately too low to result in any detectable increase in cancer or teratogenesis, but this would obviously not be the case if atomic weapons were again to be exploded in the atmosphere.

Trace elements

Traces of several elements not thought to be essential for health enter the body from food, water and air. Most appear to be harmless at low levels, but, like the essential **trace elements** (see for instance Copper) high intakes of some are known to have toxic effects. Those most likely to contaminate food are shown in Table 3.

Small amounts of arsenic occur naturally in most foods: shell fish, seaweed and hops contain more than most. Food contaminated with arsenic (for instance beer containing

glucose made from impure sulphuric acid) has caused outbreaks of poisoning in the past, but food contents are now
legally controlled and the average dietary intake is well
within the safe limit. At higher intakes, though still insufficient to cause poisoning, arsenic is a suspected carcinogen.

High intakes of cadmium cause kidney damage and
animal experiments suggest that cadmium may be a cause
of high blood pressure and possibly **heart disease**. It also
appears to be antagonistic to the essential trace elements
zinc, copper and selenium. In Japan, the disease 'Itai Itai' is
probably due to industrial contamination of water supplies
with cadmium. In Britain the average intake is well within
the probable safe limit, but people who eat much kidney and
crab would exceed the average. Other sources of cadmium
are cigarette smoke (which may be a greater hazard because
more is absorbed through the lungs than from food), soft
water left standing in galvanised pipes, and some enamelled
casseroles. A British Standard for cadmium in ceramic ware
was introduced in 1972, but casseroles bought before then
and enamelled red, yellow or orange on the inside (including
the inside of the lid) can be harmful, especially when used
for cooking acid foods.

Excessive lead intakes cause anaemia and damage the
kidneys and nervous system especially in children. Food is
the main source, but although the average intake is within
the safe limit, extra can be added during cooking. For
instance, lead dissolves out of pewter pots and from lead-
containing glazes on some earthenware casseroles. Acid
foods dissolve the most lead. Other sources of lead—water
left standing in lead pipes (soft water dissolves out more
than hard), air polluted with car exhausts (high octane fuel
contains the most lead) and lead in old paint—can be more
serious hazards. Several children die each year from eating
chips of old paint: modern paints contain much less. City
dust also contains very high levels and can be dangerous if
eaten by children.

There are several forms of mercury (quicksilver), most of

TABLE 3. HARMFUL TRACE ELEMENTS

Element	Probable safe food intake	Quantity in average British diet	Main routes of pollution	Average quantity of most polluted food necessary to exceed safe intake	Notes
Arsenic	About 20 milligrams weekly	About 2 milligrams weekly	Pesticides Impure sulphuric acid used to make food ingredients like glucose, citric acid Sometimes used as a growth promoter in the feed of poultry and pigs	Shellfish—more than 150 grams (5 oz) weekly	Legally, most foods must not contain more than 0·1 milligrams per 100 grams. Others, like spices, may contain up to 0·5 milligrams per 100 grams. Shellfish, seaweed and hops are exempt
Cadmium	About 450 micrograms weekly	About 150 micrograms weekly	Industrial wastes discharged into sea, contaminating fish Fumes from motor vehicles and factories contaminating nearby crops Superphosphate fertilisers and sewage sludge used on crops Soft water standing in galvanised pipes	Brown meat of crab—more than 50 grams (2 oz) weekly Winkles, shrimps, mussels from Bristol Channel—more than 100 grams (3½ oz) weekly Other shellfish and white crab meat—more than 250 grams (½ lb) weekly	Bristol channel is only significantly polluted estuary

Element	Probable safe food intake	Quantity in average British diet	Main routes of pollution	Average quantity of most polluted food necessary to exceed safe intake	Notes
Lead	About 3 milligrams weekly from food and water	About 1·5 milligrams weekly from food and 0·15 milligrams from water	Insecticides Sewage sludge used as fertiliser Fumes from motor vehicles and factories contaminating nearby crops Lead in solder used for sealing canned foods	Vegetables grown by roadsides—outer leaves, more than 150 grams (5 oz) weekly—inner leaves, more than 1 kilogram (2 lb weekly)	Legally most foods should not contain more than 0·2 milligrams per 100 grams. Canned foods may contain up to 0·5 milligrams and fresh apples and pears 0·3 milligrams per 100 grams
Methyl Mercury	About 250 micrograms weekly	About 70 micrograms weekly	Industrial discharges and natural erosion of rocks into sea, contaminating fish Pesticides	Fish—more than ½ kilogram (1 lb) weekly of pike and fish from some polluted estuaries Tuna—more than 1½ kilograms (3 lb) weekly Shellfish—more than 2 kilograms weekly	Thames, Mersey and Lune Deep are polluted

which are toxic in excess, but methyl mercury is the most dangerous. It damages the brain and nervous system irreversibly and is probably a carcinogen, mutagen and teratogen. Since 1953 at least 46 people have died in the Minamata Bay area of Japan from eating polluted fish. Mercury was discharged from local factories into the sea where it is converted by bacteria into methyl mercury. The fish were found to contain up to 50 micrograms per gram: only about 5 grams would have exceeded the safe limit. In Britain, nearly all fish landed is relatively unpolluted and the quantity of mercury in the average diet is well within the probable safe limit. Pike, tuna and shellfish contain more than other fish and excess intakes should be avoided (see Table 3). However, although discharges into the sea are now being brought under control, all fish (including white) is polluted in some estuaries, like the Thames, Mersey and Lune Deep. People eating local catches could be at risk.

Methyl mercury is also sometimes used as a fungicide for seeds. Although the mercury left in the soil adds to general pollution, plants absorb little and the harvest is not affected. The main danger to humans is when the fungicide-dressed seed is actually eaten—as happened in Iraq in 1973 when over 450 people died. There have been similar disasters in Pakistan and Guatemala. Birds eating the grain are also killed for this reason, Sweden banned their use in 1966. Little is used in Britain.

Miscellaneous

Excessive use of fertilisers can lead to high **nitrate** levels in crops. See also Antibiotics. Many other industrial chemicals —including the very toxic polychlorbiphenyls, which, like DDT, are recycled in the environment—can pollute food. For instance, wood may be preserved with chlorophenols and these may contaminate food when the wood is used as poultry litter, or made into paper for packaging. Some chlorophenols have a disinfectant taste, even at almost undetectable levels. If wood is burnt, chlorodioxins, which are extremely toxic and suspected carcinogens, are formed from chlorophenols and contaminate food in contact with the smoke in the home, but even food smoked with fresh wood is not harmless. Smoke contains polycyclic hydrocarbons—the potent carcinogens in cigarette smoke and motor vehicle exhausts—which are deposited on smoked and barbecued food. Icelanders, who eat a lot of home smoked food, have one of the highest rates of stomach cancer.

Poisons in plants and moulds

Some mushrooms and toadstools are well known poisons and a variety of plants contain potential poisons. Toxins include thujone in wormwood (used in absinthe and vermouth) which in excess may cause convulsions; myristin in nutmeg which in excess can cause coma; solanine in potatoes; oxalic acid in spinach; arsenic in hops; and cyanide in bitter almonds and slivovitz. Chick peas also contain a neurotoxin (damaging to the nervous system) and in excess cause lathyrism—a disease which often results in spastic paraplegia—and sassafras, used in some herbal teas, contains a liver carcinogen.

Perhaps the most insidious and dangerous natural toxins are made from fungi. Ergotism, caused by the fungus ergot growing on rye, is well known, but only recently has it been realised that some seemingly harmless mouldy food contains carcinogens. One of the aflatoxins, formed by the mould *Aspergillus flavus* growing on poorly stored groundnuts and other food is a potent cause of liver cancer, at least for animals and almost certainly man. It is probably worthwhile to avoid all mouldy food in the home: many other mycotoxins are suspected carcinogens. The USA Food and Drugs Authority advises throwing away all mouldy food: trimming it off is not a sufficient precaution because the 'roots' may penetrate into the food.

Food poisoning

Food is an important potential source of infections, for instance, watercress can be infected with liver fluke, and pork (which should always be thoroughly cooked) with tape- and roundworms. In Britain, these risks are however slight compared with those from food poisoning.

Every year, at least 10,000 people in Britain succumb to food poisoning. Bacteria are usually responsible, though the cause cannot always be traced and some outbreaks may be due to viruses.

In warm, moist food, one bacterium can multiply to result in thousands of millions in a day. Most bacteria produce 'off' flavours as they multiply, when the food would not be eaten, but food poisoning bacteria are dangerous because they increase without affecting the apparent wholesomeness of the food. Two to 36 hours after eating the food, the unsuspecting consumer suffers from stomach cramps, vomiting and diarrhoea. Meat is the offending food in at least three-quarters of cases and nearly all outbreaks occur when food is carelessly prepared in bulk (for instance in canteens, hospital kitchens, restaurants and other institutions). In the home, food poisoning can be avoided if the precautions given in Table 4 are followed.

There are many food poisoning bacteria, but salmonellas are responsible in at least half of traceable cases. Raw milk, eggs, meat and poultry can be contaminated on the farm or by droppings from pets or pests. They are destroyed by heat and freshly cooked food is theoretically safe. However meat (especially poultry, which is very likely to be contaminated) not properly thawed before cooking is dangerous because the inner part may not have reached a sufficiently high temperature to kill all the salmonella. In Sweden in 1953, 105 people died from salmonella in meat balls that had been inadequately cooked. In the kitchen, salmonella can be transferred from one food to another by hands and equipment. Some outbreaks have been caused by portions of

TABLE 4. PRECAUTIONS AGAINST FOOD POISONING

Precaution	Reason
(1) High standard of hygiene (scrupulously clean hands, kitchen and equipment)	Unwashed hands, dirty equipment, flies and other pests (attracted by dirt and rubbish) can contaminate food. 2 and 3 are as important because it is difficult to eradicate food poisoning bacteria completely from hands and some equipment (like wooden chopping boards)
(2) Food must always be cooled quickly after cooking, and always immediately stored in a refrigerator. Cold meat, pies, cream cakes, trifles, mayonnaise etc., which are not cooked before eating must never be left to stand in a warm place	Food poisoning bacteria cannot multiply below 4°C, the temperature of a well maintained refrigerator
(3) Food must always be cooked thoroughly	Heat kills most food poisoning bacteria
(4) Meat should always be completely thawed before cooking	Centre of frozen meat may not reach a sufficiently high temperature to kill bacteria
(5) Avoid reheating meat and casseroles	Spores may survive cooking, germinate and multiply whilst the food is cooling or being reheated. If absolutely essential to reheat meat, it must be cooled quickly, stored in a refrigerator, and quickly and thoroughly reheated
(6) Keep raw and cooked foods separate, and avoid using the same surface for preparing raw and cooked food, unless thoroughly washed in very hot water and detergent	Avoids recontamination of cooked food
7) Rehydrated and food in opened cans should be treated as if it were fresh	Bacteria cannot multiply in dry food, and the sterile contents of cans are protected by the can. Once opened or rehydrated, these foods are as liable to contamination as other food
(8) Meat should never be bottled or canned in the home	Danger of botulism
(9) Never taste suspect food	Food poisoning bacteria do not produce off flavours

chicken left out of the refrigerator after being cut up on a surface that had previously been used for preparing the raw chicken for the oven.

Clostridium welchii is responsible for up to a third of traceable cases. The spores of these bacteria are resistant to cooking and are found in dust and air: large joints of meat left to cool uncovered and partially reheated later may contain dangerous numbers. Meat—especially reheated meat and pies—is the usual source of infection.

About a further 5% of traceable cases are caused by staphylococci, but it is the toxins, and not the bacteria themselves, that cause poisoning. Most people have staphylococci in their noses and the bacteria are transferred to hair and hands. Handling food with unwashed hands, and sneezing, coughing and smoking over food can deposit the bacteria on it. They multiply—forming toxins—unless the food is refrigerated. The bacteria are destroyed by heat, but the toxins can withstand boiling for up to an hour—thus even thoroughly cooked food can be harmful if it has not been kept in a refrigerator. Foods that are not cooked after preparation, like trifles, custards, cream cakes and cold meat, are however, the most liable to be toxic.

Clostridium botulinum—the cause of botulism—can only multiply in air-free non-acid foods, like canned and bottled meats and some vegetables, and smoked food. **Nitrite** added to cured and smoked meat inhibits its growth, and sea **fish** is unlikely to be contaminated. However, there is a risk from smoked trout unless it is kept refrigerated. All commercially canned food is heated to a temperature sufficiently high to kill spore forms but home canned and bottled meat is dangerous because it is difficult to reach the necessary temperature with domestic equipment. Spores which escape sterilisation germinate and multiply whilst the can is stored, forming a lethal nerve toxin. The consumer is severely ill within 3 days. Botulism is fortunately rare in Britain: it is fatal in about three-quarters of cases.

POLYPHOSPHATES

Legally permitted miscellaneous **additives** in food. They are classed as acids, and retard the growth of bacteria and moulds that would otherwise grow in foods like meat and fish. More importantly to the food industry, they prevent the normal loss of water that occurs on cooking, when proteins shrink, squeezing water out of food. The meat is more succulent, but also contains less nutrients, weight-for-weight than traditional products. The effect of polyphosphates is most marked in modern hams.

POLYSACCHARIDES

Long chains of simple sugars (**monosaccharides**). Some, like **starch, dextrin** and **glycogen**, can be split to their constituent sugar, glucose, during digestion which is then absorbed into the blood stream and used for energy. Others, like **cellulose, edible gums, agar, alginates, pectin** and **inulin** cannot be digested and contribute to dietary **fibre**.

POTASSIUM

An **electrolyte** in body fluids, necessary for the proper functioning of cells, including nerves. Adults contain between 100 and 150 grams ($3\frac{1}{2}$ to 5 oz), most of which is inside cells. It cannot be stored, and between $\frac{1}{2}$ to 1 gram is lost every day in the urine. This loss must be replaced by potassium absorbed from food.

Nearly all foods—except sugar, fats, and spirits—contain potassium. Rich sources are instant coffee, **extracts**, wheat germ, black treacle and dried fruits. **Nuts** contain 100 to 240 milligrams per 25 grams (1 oz); meat, fish, wholemeal bread and potatoes 70 to 130 milligrams; white bread, cheese, milk and eggs 30 to 50 milligrams. (One milligram is equal to one thousandth of a gram.) Vegetables and fruits contain between 20 and 300 milligrams per 25 grams. Boiling of vegetables reduces their potassium content by half, but the mineral can be recovered in the cooking water. None is lost on baking or frying. Instant coffee contains 250 milligrams per cup, ground coffee 150 milligrams per cup, tea 25 milligrams per cup. Beer contains 100 milligrams per $\frac{1}{2}$ pint, soft drinks usually nil.

Potassium deficiency is rare in healthy adults and children. Nearly all of the 2 to 6 grams of potassium in the average diet is absorbed into the blood stream and amply replaces body losses. If deficiency is likely—when there are increased losses from the body, for example in chronic diarrhoea, vomiting, diabetes, some kidney diseases and from diuretics (drugs prescribed to increase the volume of urine)—the losses are counteracted by extra potassium, given as pills or intravenous fluids. However, excessive use of purgatives can also deplete the body. The elderly, who frequently eat less than average quantities of potassium, are particularly at risk from the resultant weakness, mental confusion and—in severe cases—heart attack. Potatoes (each portion contains about $\frac{1}{2}$ gram) and instant coffee are a convenient way of adding potassium to the diet.

In health, a high potassium diet is not harmful—the excess is eliminated by the kidneys, but in chronic kidney disease unwanted potassium accumulates in the blood stream, and can result in a heart attack.

POTATOES

The potato is perhaps the most unfairly maligned food. When first introduced into Europe it was thought to have weakening properties (or even be a cause of leprosy) and now it has an undeserved reputation as a purely fattening food. In fact, potatoes supply vitamin C, minerals, most B vitamins and protein.

Compared with 210 kilo**calories** in 50 grams (approximately 2 oz) of cheddar cheese, 50 grams of boiled potato only contains 40 kilocalories. 80% of its weight is (non-fattening) water. If no fat is added, they are suitable for inclusion, in measured amounts, in **slimming diets**. However, chips, roast potatoes and crisps contain less water and added **fat**. Weight-for-weight, chips contain three times more **energy** than boiled potatoes.

About 20% of the weight of boiled potatoes is **starch** but it is a comparatively poor source of **carbohydrate** compared with plain biscuits (75%) or sugar (100%). 50 grams of potato contains 10 grams of carbohydrate—about the same as an apple. Potatoes also contain **fibre** and, perhaps surprisingly, small quantities of good quality **protein**.

Potatoes are an important source of **vitamin C** (especially in winter), supplying a quarter of the total in the average British diet. They are rich in **potassium** (fried potatoes contain more than meat) and supply moderate quantities of other **minerals.** They are lacking in vitamin B_{12} (like other vegetable foods), poor sources of vitamin B_2 (**riboflavin**), but are useful sources of other **vitamins B**. Potatoes contain **vitamin E** (crisps fried in oil are good sources) and yellow (but not white) sweet potatoes are sources of **carotene** (pro vitamin A).

Fresh potatoes are living, and vitamin C declines during storage. August and new potatoes contain about 30 milligrams of vitamin C per 100 grams, but by March the level has fallen to about 10 milligrams. Old potatoes eaten after May may contain no vitamin C. Bruised potatoes contain less than sound ones, and poor storage conditions will hasten its destruction. Ideally they should be kept in a dark, dry, cool (between 5 and 10°C) place. In light, they turn green and sprout, forming solanine, which can be toxic. The eyes and green parts should be cut out. Potatoes should be taken out of plastic bags, otherwise they become moist and start to rot. They quickly lose weight and vitamin C in warm places, but may become sweet (the starch is converted into sugar) if kept too cool, for example in a refrigerator.

Up to half of vitamin C and potassium can be lost when

potatoes are boiled; less is lost if they are cooked in a small quantity of water for as short a time as possible. Stewed potatoes (for instance in soup) contain none. No potassium is lost during baking, frying and pressure cooking and only about a third of vitamin C is destroyed. After cooking,

AVERAGE NUTRIENTS IN POTATOES

Nutrient	in 150 grams (5¼ oz) boiled potatoes	in 150 grams roast potatoes	in 100 grams chips	in 25 grams crisps
Energy (kilojoules)	505	775	1005	585
(kilocalories)	120	185	240	140
Protein (grams)	2	4¼	3¾	1½
Carbohydrate (grams)	30	41	37	12
Fat (grams)	Trace	1½	9	9½
Potassium (milligrams)	490	1120	1020	340
Sodium (milligrams) (1)	5	15	10	10
Calcium (milligrams)	5	15	15	10
Magnesium (milligrams)	25	50	45	10
Iron (milligrams)	0·7	1·5	1·4	1
Phosphorus (milligrams)	45	80	70	30
Vitamin A, D	0	0	0	0
Vitamin C (milligrams)	6 to 30	7 to 36	5 to 22	—
Vitamin E (milligrams, α tocopherol)	0·07[a]	b	b	b
B vitamins				
thiamin (milligrams) (2)	0·12	0·15	0·1	—
riboflavin (milligrams)	0·05	0·06	0·04	—
nicotinic acid (milligram equivalents)	1·8	3	2·2	—
pantothenic acid (milligrams)	0·5[a]	—	—	—
pyridoxine (milligrams)	0·3[a]	—	—	—
folic acid (micrograms)	15[a]	—	—	—
biotin (micrograms)	0·15[a]	—	—	—
vitamin B_{12}	0	—	—	—

(1) no salt added in cooking or processing
(2) half the thiamin may be destroyed if potatoes treated with sulphur dioxide
Instant potatoes may contain virtually none
[a] – value for raw potato
[b] – content depends on oil used for frying. Crisps are usually good sources
— – no figures available

keeping potatoes hot further reduces vitamin C (by about half in two hours) but there is less destruction in boiled, rather than mashed, potato. Old potatoes eaten in late spring in a canteen or restaurant are likely to contain no vitamin C. Peeled potatoes should not be left to soak, otherwise vitamin C will leach out into the soaking water.

About two-thirds of vitamin C is lost when potatoes are blanched but there is no loss on freezing. Canned potatoes are likely to contain as much or more vitamin C than fresh because they are processed when the vitamin content is maximal and losses are no greater than those incurred in cooking. Dehydration however can destroy all vitamin C: instant potatoes with added vitamin C (declared on the label) should be chosen in preference to others. Sulphur dioxide (a **preservative**) is added to ready prepared potatoes and chips to prevent browning: it preserves vitamin C, but also destroys **thiamin** (vitamin B_1).

The potato did not become established in England until the beginning of the nineteenth century. At that time the cereal harvest failed and the value of the potato was recognised. After its acceptance as a staple food, **scurvy** declined in England. Now that the era of cheap imported food is over, potatoes will no doubt again become increasingly important in British diets. They are easy to grow and provide more food energy per acre than cereals. Their relative cheapness and versatility (there are over 100 different ways of cooking potatoes) make them an ideal base for balanced hot filling meals. Their lack of vitamin A is easily overcome by adding margarine (or butter) and a sprinkling of cheese adds extra protein, and vitamins B_2 and B_{12}.

The suspicion, first voiced in 1972, that blighted potatoes cause spina bifida has since been disproved.

PRESERVATIVES

Substances added to food to retard decay. Like other ways of preserving food (deep freezing, canning, dehydration) they inhibit the multiplication of micro-organisms (bacteria, yeasts and moulds) and may also inactivate **enzymes** in food which would otherwise cause deterioration of flavour and colour.

Permitted preservatives

The Preservatives in Food Regulations 1974 permit the
addition of 7 groups of chemical preservatives, in specified
amounts, to about 50 foods: see table below.

Sulphur dioxide (see Sulphur) is the most commonly used
preservative. It is an effective steriliser (kills bacteria) and is
especially useful in fruits and vegetables because it prevents
browning (see Colour) and preserves vitamin C. However,
it destroys **thiamin** (vitamin B_1) and is only permitted in
meat products (an important source of thiamin) which
would otherwise be a dangerous source of food poisoning
bacteria (see Poisons in food).

Benzoic acid is found naturally in loganberries and prunes.

Food group	Food allowed to contain preservative	Permitted preservative
Meat	Bacon, ham, pickled meat	Nitrite
	Sausages, sausage meat, or any other mixture of raw meat and cereal, like beefburgers, hamburgers	Sulphur dioxide
Vegetables and fruit	Raw peeled potatoes or chips, any dehydrated vegetable, candied peel, dried fruit, jam	Sulphur dioxide
	Fruit juices, fruit pulp, tomato purée or pulp, pickles, sauces	Sulphur dioxide or benzoic acid
	Bananas or citrus fruit (on the skin)	Phenyls
Beverages	Beer, cider or perry	Sulphur dioxide
	Wine	Sulphur dioxide or sorbic acid
	Soft drinks, glucose drinks	Sulphur dioxide or benzoic acid
	Instant coffee or extract	Sulphur dioxide
	Drinking chocolate, tea extract	Benzoic acid
Dairy	All cheese	Nisin and sorbic acid
	Clotted cream	Nisin
	Any cheese except cheddar, cheshire, soft cheese	Nitrite
Flour	Bread	Proprionic acid
	Cakes, biscuits, pastry	Proprionic acid or sorbic acid
Miscellaneous	Any canned food	Nisin
	Gelatine, ginger, marzipan, caramel, vinegar, sugar, glucose syrup	Sulphur dioxide
	Colours	Benzoic acid or sorbic acid
	Flavours	Sulphur dioxide or benzoic acid

It is allowed in six foods, and as an alternative to sulphur dioxide in some others. Proprionic acid and sorbic acid also occur naturally in food. They prevent the growth of moulds and are converted to harmless substances in the body. Nisin is an **antibiotic** found naturally in cheese and is split to its constituent **amino acids** during **digestion**. Citrus fruits and bananas may have phenyls on the skin, to prevent mould: when used in cooking the skin of citrus fruit should be well washed. See also Nitrite.

Added preservatives must be declared on the **label** of most foods. The regulations also allow small amounts of formaldehyde (used to make some wrappings) to contaminate food, but do not control sprout inhibitors (used on potatoes) and preservatives added whilst the food is held in bulk storage (for example fungicides, insecticides). These are under the general control of the Food and Drugs Act—see Additives.

'Natural' preservatives

Most micro-organisms cannot multiply in acid or very concentrated or dry foods. **Acetic acid** (in vinegar) is used to pickle vegetables, and **lactic acid** (from bacteria) in sauerkraut and yoghurt. Lactic acid is also formed in meat after slaughter. When used in sufficient quantities, **sugar** and **salt** are effective preservatives (for example in salted meat and fish, preserves, and dried fruits). Other natural preservatives are alcohol and substances in some spices (like cloves), hops and smoke.

There is less destruction of vitamins when acids, sugar and salt are added to canned foods (with these added inhibitors less heat for sterilising the can is necessary). For example, there is severe destruction of B vitamins in evaporated milk but only moderate destruction in condensed.

PRESERVES

Jam and marmalade making is the traditional way of preserving excess fruit. The fruit is sterilised by boiling, and sugar (sucrose) acts as a preservative. Both sugar and pectin are necessary for the jam to set. Pectin is degraded as fruits ripen: overripe fruit (or those naturally low) require added pectin or are combined with a fruit high in pectin, like apples or plums.

About 70% of the weight of preserves is sugar. 25 grams of jam or marmalade doubles the energy and carbohydrate of a thin (300 gram) slice of bread, see table below. Yellow fruit (like apricots, marmalade) preserves contain carotene. Although the B vitamins are largely destroyed, some vitamin C in fruits which are rich sources (like blackcurrants) survives the boiling process.

Jams and marmalade are manufactured from slightly imperfect fruit unsuitable for canning or freezing. Legally they

AVERAGE NUTRIENTS IN 25 GRAMS JAM AND MARMALADE

Nutrient	Jam	Marmalade
Energy (kilojoules)	275	275
(kilocalories)	65	65
Carbohydrate (grams)	17¼	17¼
Protein (grams)	0·1	Trace
Fat	0	0
Sodium (milligrams)	5	5
Potassium (milligrams)	25	10
Calcium (milligrams)	5	10
Magnesium (milligrams)	5	1
Iron (milligrams)	0·3	0·15
Phosphorus (milligrams)	5	5
Other minerals	Small amounts	
Vitamin A, B (equivalents)	Trace	4
Vitamin C (milligrams)	3[a]	3
Vitamin E	—	—
B Vitamins	Trace	Trace

a – blackcurrant jam contains 12 milligrams
— – no figures available

must contain not less than 65% of sugar and fruit solids, and at least 20% of fruit in marmalade (30% in jam) must be used in the recipe. Cheaper preserves may be made from frozen fruit pulp, with added sulphur dioxide (a preservative). Jams may also contain added **colour** (dark marmalades may be coloured with caramel), flavours and fruit acids (see Miscellaneous **additives**). Small quantities of **mineral hydrocarbon** and **antioxidant**, added to some fruits, may also remain in the jam.

PROTEIN

Protein is part of the structure of each of the millions of cells in the human body. It is constantly being broken down and remade; cells are constantly dying, and in most cases, being replaced. Some of the proteins are discarded in the daily turnover, and an equivalent amount must be provided in the diet. Babies, children and adolescents need more protein in proportion to their weight than adults because they are growing and making many new cells. Pregnant and nursing mothers also need more than usual to ensure adequate supplies for the growing child.

No natural food is pure protein. Even low-fat soya flour—the most concentrated natural source—only contains half its weight (50%) as protein. Meat, fish, cheese, eggs, most nuts (particularly peanuts), dried peas, beans and lentils however contain a high proportion—between 10 and 30%. Cereals contain 10%. Fresh milk, because it is mainly water, contains a low percentage but it is taken in larger quantities and contributes a significant quantity to the diet. Dried skim milk contains 34% of protein. Other vegetables contain varying amounts, but never more than 5%. Fruits contain very little; sugar, fats and alcoholic spirits none.

Most adults eat about 70 grams of protein a day, of which nearly two-thirds (60%) is derived from meat, fish, cheese, eggs and milk. Cereals, including bread, contribute another quarter (27%), and vegetables a further 10%. The table below demonstrates the approximate protein content of some average portions of food.

Type of food	Grams of protein *in an* average portion	
Fish	18	100 grams (3½ oz) steamed cod
Meat and poultry	13	50 grams lean roast beef
Cheese	12	50 grams (two thick slices) cheddar
Milk	10	300 millilitres (½ pint)
Egg	7	60 grams (1 standard)
Nuts	14	50 grams peanuts
Pulses	8	150 grams (1 small tin) baked beans
Cereals	8	100 grams (3 thin slices) bread
Vegetables	2	150 grams (moderate portion) boiled potatoes
Fruit	½	One medium apple

In general, proteins from animal sources (meat, fish, cheese, milk and eggs) are more efficiently used to replace daily losses from the body and form new cells than vegetable proteins. Most animal proteins are therefore of 'better quality' than vegetable proteins. For instance, the 7 grams of good quality protein in an egg can all be used for replacement, but only half the 7 grams in 3 slices of bread. There are exceptions to this rule: gelatine is hardly used at all, and mixtures of vegetables (like rice and lentils) are more efficiently used. (For more detailed explanation see Amino acids.)

Proteins are less efficiently used if there is a high ratio of protein to **energy** in the diet. A high protein diet, containing a lot of meat, etc, is part of the treatment of several medical conditions, but when eaten by healthy adults the excess protein is turned into **fat** or **glucose** and used for energy. Although high protein diets are not harmful for adults—because the excess unwanted waste products are filtered out of the blood by the kidney into the urine—they are nutritionally wasteful. However, high protein foods are good sources of other nutrients (particularly B **vitamins** and **iron**) and meals that contain a moderate portion of high protein food, together with cereals and fruit and vegetables, will ensure a well **balanced diet**.

Structure of proteins

There are many different proteins but all are composed of long chains of amino acids. Each protein contains its own specific number (sometimes thousands) of amino acids in a specific sequence in the chain. The chain is folded and twisted into complex structures, dictated by the amino acid sequence, which in turn is determined by the individual genetic template (DNA) in the nucleus of cells. Most foods contain several different proteins—for instance there are at least 5 different proteins in meat, and 3 in milk—but it is the total sum of individual amino acids that determines the protein quality of a food.

Heat alters the structure of proteins, causing the chains to unfold partially. Acids, such as lemon juice, can have the same but limited effect. Changes in the protein structure, coagulation, occur when most food proteins are cooked. For instance, the proteins in raw egg white are dissolved in water, but when cooked they are coagulated and made insoluble. The white hardens and becomes opaque. The temperature at which proteins coagulate varies: egg white proteins coagulate at 60°C, egg yolk proteins at 65°C, and milk proteins at 100°C. If the heating is prolonged, proteins are toughened and shrink, squeezing water out of the food. Thus overcooked scrambled egg, meat and fish is dry and tough.

Usually cooking and processing do not affect the nutritional value of protein, and light cooking may improve digestibility (the amino acids in an unfolded (coagulated) chain are more accessible to digestive enzymes). Tough overcooked proteins are however less digestible, and toasting and puffing of breakfast cereals (like cornflakes) may reduce the quality of their protein. Amino acids, particularly **lysine**, are destroyed by dry heat. Lysine is also inactivated when sugars react with it, forming a brown coloured substance, in foods like dried milk.

Protein in the human body

A 70-kilogram (11-stone) man contains about 13 kilograms
(2 stones) of protein, distributed, in different types, throughout all his cells. Nearly all parts of the body contain collagen,
a protein especially abundant in bones, skin and muscles.
Keratin is in nails and hair, elastin in tendons and arteries.
In muscles, which contain 40% of the total body protein,
there are two special proteins able to move together and
apart when muscles contract and relax. The thousands of
enzymes and some of the hormones—like insulin—are also
proteins. In blood, proteins like the oxygen carrying red
pigment haemoglobin, are used to carry nutrients from one
part of the body to another. Other blood proteins are involved in protecting the body against infection. All these
proteins are continually replaced at different rates: enzymes
are exchanged very quickly, but bone collagen very slowly.
Most of the constituent amino acids from the dismantled
proteins are recycled, but some are discarded every day and
must be replaced by amino acids derived from proteins in
food.

All proteins in food, from both animal and plant sources,
except the very tightly coiled keratin, are split to their component amino acids by enzymes during **digestion** and transported, in the blood stream, to the liver. About 98% of
protein is digested, but less when the diet contains a lot of
roughage (**fibre**). Tiny quantities of protein may be absorbed intact; this is an advantage for babies fed on mothers'
milk since they can absorb antibodies in this way, but foreign
food proteins in the blood stream can cause allergies.
Throughout the day, amino acids are released into the blood
stream and taken up by the cells and used for replacement
and growth.

Protein requirements

The minimum protein requirement for adults depends on
the number of active cells in the body. Men have more

muscle cells, and therefore require slightly more than
women and the elderly. Allowing for individual variation,
and for the fact that only 70% of the protein in a British
mixed diet is used to make new body protein, the minimum
protein requirements are 45 grams per day for a man weighing
65 kilograms (9½ stones), and 38 grams for a woman weighing 55 kilograms (8½ stones). People with a tall or heavy
frame (but not overweight since fat cells contain very little
protein) need more than this. Most people in Britain eat at
least twice this amount, and would find a diet containing
so little protein unpalatable. Additionally, such low protein
diets may incur the risk of a B vitamin (and especially in
women) iron deficiency. The **recommended intakes** are therefore higher than the minimum, and depend on the ratio of
protein to kilocalories, varying from 55 to 90 grams per day
for adult men and women. (See Appendix.)

Children and adolescents need protein both for maintenance and growth. For instance, a child who puts on 5 kilograms (11 lbs) in a year between the age of 12 and 13 would
need an extra 4 to 4½ grams of protein each day above the
requirement for replacement of protein to allow formation
of new cells. During pregnancy, mothers need about an extra
5 grams daily in the last 6 months above the usual requirements. See Appendix for total requirements.

For some as yet unexplained reason, more proteins are
lost from the body in times of stress like during recovery
from operation, burns, or accidents. The requirements for
protein are therefore greater during convalescence when the
'stress' reaction ceases, and losses can be replaced.

Excesses/deficiency of protein in the diet

Protein eaten in excess of the daily requirement cannot be
stored. The potentially toxic **nitrogen** is removed from amino
acids and converted to urea and other relatively harmless
substances, later filtered out of the blood by the kidneys
and excreted into the urine. The other part of amino acids
is converted into fat or glucose, and stored or used immedi-

ately for energy. Protein can thus be as fattening as carbohydrates or fats.

Although a high protein diet is not harmful for adults too
much protein is harmful for babies. The kidneys are immature during the first few months of life and unless infants
are given sufficient water, the extra urea formed from excess
protein is not excreted and accumulates in the blood stream,
making the baby thirsty and nauseated.

The body can become depleted of protein both by diets
low in protein (for instance in alcoholism) and in **starvation**,
when amino acids are diverted away from repair and used
for energy. Children are more readily affected—see Kwashiorkor and Marasmus. All bodily processes are interrupted
but initially non-essential protein replacement, in hair and
nails, and growth are checked. Later muscles are wasted
away to supply amino acids for more vital proteins like
enzymes; the liver accumulates fat and blood proteins are
not replaced. Eventually the heart and digestive tract degenerate, and death results from heart or liver failure.

Protein and special diets

The liver has a central role in protein metabolism, forming,
for example, blood proteins and converting unwanted amino
acids into urea. The kidneys are responsible for conserving
blood proteins and eliminating urea and other waste products from the body. The level of blood proteins may fall in
the early stages of liver or kidney disease: water then seeps
out of the blood stream and body fluids accumulate, causing
oedema (waterlogging). Extra protein may be necessary and,
as the diets are sometimes restricted in **sodium**, may be
given in the form of special high protein, low salt drinks.

In later stages, low protein high energy diets are necessary
to minimise the work of the diseased liver or kidneys and
to avoid accumulation of urea and other waste products
which are toxic in excess. To meet the bodily needs for
essential amino acids, most of the small quantity allowed
is taken as carefully calculated and weighed portions of

animal protein (eggs, milk, meat or fish). In very low protein diets (20 grams per day) used for kidney failure, cereal, nut, pulse and other sources of vegetable protein are almost entirely excluded. **Gluten**-free flour (wheat starch) is used in place of ordinary for bread, cakes, pastries and biscuits. The benefits of low protein diets are not felt, if the diet contains insufficient energy. Plenty of butter, double cream, other fats, sugar and preserves are essential, otherwise wasting (and production of urea) continue. Special glucose drinks and other low protein, high calorie foods are available on prescription when necessary.

PULSES—also called legumes

Peas, beans and lentils, seeds of the leguminosa family of plants, are richer in **protein** and the B vitamins **thiamin** and **nicotinic acid** than other **vegetables**, and contain more **carbohydrate** and **energy**. Like other vegetables, pulses supply other B vitamins, **minerals** and **fibre** and all (except soya beans) are fat free. String beans, runner beans and sugar peas (mange-toute) are eaten in the pod and have a similar composition to other green vegetables.

A portion (150 grams) of baked beans contains as much protein as an egg, but when eaten alone, less than half is used to make new body protein for growth and repair. Although most pulse protein is of poorer quality than animal protein, when eaten with cereals (like rice, bread) the mixed proteins supplement each other and are used with greater efficiency. Pulses are relatively lacking in the essential **amino acid** methionine (but are good sources of lysine) and cereals are relatively poor in lysine (but are good sources of methionine). Lentils are poorer in methionine than other pulses.

The nutritional quality of the soya bean is superior to other pulses. Its **iron** is more readily absorbed into the blood stream and its protein is of better quality. It contains more methionine and, when eaten alone, nearly as much

AVERAGE NUTRIENTS IN 100 GRAMS (3½ OZ) OF SOME PULSES

Nutrient	Baked beans	Broad beans (boiled)	Butter beans (boiled)	Haricot beans (boiled)	Lentils (boiled)	Fresh/frozen peas (boiled)	Dried peas (boiled)	Split peas (boiled)	Tinned peas (uncooked)	Soya beans (dried, uncooked)
Energy (kilojoules)	390	180	390	375	405	205	420	485	360	1605
(kilocalories)	95	45	95	90	95	50	100	115	85	380
Protein (grams)	6	4	7	6¼	7	5	7	8¼	6	35
Fat (grams)	½	0	0	0	0	0	0	0	0	18
Carbohydrate (grams)	17¼	7	17	16¼	18¼	7¼	19	22	16¼	20
Sodium (milligrams)	590a	20	15	15	10	Trace	15	15	260a	—
Potassium (milligrams)	345	235	400	320	215	175	265	270	200	—
Calcium (milligrams)	60	20	20	65	10	15	25	10	25	200
Magnesium (milligrams)	35	25	35	45	20	20	30	30	25	—
Iron (milligrams)	2·1	1·0	1·7	2·5	2·2	1·2	1·4	1·7	1·9	7
Phosphorus (milligrams)	185	100	85	120	80	85	115	120	170	—
Vitamin A (microgram equivalents)	50	20	0	0	Trace	50	15	—	50	Trace
Vitamin D	0	—	0	0	0	0	0	0	0	0
Vitamin E (milligrams α tocopherol)	—	—	—	—	—	0·5b	—	—	0–10d	0
Vitamin C (milligrams)	3	15	0	0	0	15	0	0	0	0
B vitamins thiamin (milligrams)	0·06	0·28b	0·45b	0·45b	0·5b	0·25	0·11c	0·7b	0·06–0·12	1·1
riboflavin (milligrams)	0·04	0·04	0·13b	0·13b	0·25b	0·11	0·07	0·2b	0·04–0·07	0·3

	Baked beans	Broad beans (boiled)	Butter beans (boiled)	Haricot beans (boiled)	Lentils (boiled)	Fresh/frozen peas (boiled)	Dried peas (boiled)	Split peas (boiled)	Tinned peas (uncooked)	Soya beans (dried uncooked)
nicotinic acid (milligram equivalents)	1·5	3·6	3·7b	3·7b	6·3b	2·3	6·2b	6·5b	1·6d	11
pyridoxine (milligrams)	—	—	—	—	—	0·16b	—	0·3b	2·3	—
pantothenic acid (milligrams)	—	5·4b	—	—	—	1·5b	—	2b	0·05	—
biotin (micrograms)	—	3·2b	—	—	—	0·5b	—	—	0·15	—
folic acid (micrograms)	—	—	—	—	—	5b	—	—	2	—
vitamin B$_{12}$	0	0	0	0	0	0	0	0	0	0

— – no figures available; a – salt added in processing; b – value for raw food; c – little thiamin if preservative added; d – higher figure for garden peas, lower for processed soya protein as cheese protein is used to make new body protein. Soya is an important ingredient in traditional Chinese cuisine: the beans are eaten whole or processed into sauce (like soy sauce) or into soya milk or cheese. Modern varieties of the soya bean contain 20% fat (originally soya was fat free) and are an important crop in the USA. The residue left after oil has been extracted is used for soya flour (added to many foods), **novel proteins** and cattle fodder. As yet, soya is not cultivated in Britain.

Fresh and frozen pulses—peas and broad beans contain vitamins A and C, and 4 to 5 times more thiamin than cooked beef. Vitamin C declines after harvesting: freshly picked peas contain more than those bought from a shop. There are no further losses in storage if peas are kept in a refrigerator, but up to half can be lost over a week at room temperature. About 40% of vitamin C is lost when fresh peas are cooked.

Freezing has no effect on vitamin C, A or thiamin, but some vitamin C leaches out during blanching (dipping in hot water to inactivate **enzymes** which would otherwise cause loss of vitamin C and deterioration of flavour and colour). If they are not thawed before cooking, cooked frozen peas may contain as much vitamin C as cooked fresh peas bought from a shop: peas are frozen immediately after harvesting (when their vitamin C content is maximal) and are cooked for a shorter time than fresh peas (when less vitamin C is lost). Frozen peas lose little vitamin C in a well maintained deep freeze (about 10% over a year), but there can be considerable losses at higher temperatures (25% in four months, and 80% in a year, in a freezing compartment of a refrigerator).

Canned and dehydrated pulses—dehydrated pulses are treated with sulphur dioxide (a preservative) and retain nearly all of their vitamins C and A, but lose most of their thiamin. Although there are losses of vitamin C in canning, canned *garden* peas require little cooking (and are also canned immediately after harvesting): cooked canned garden peas may contain only slightly less vitamin C than cooked fresh bought peas. Canned *processed* peas however are dried before canning and may contain no vitamin C. About half the thiamin is destroyed during canning, but there are no losses of vitamin A. Little vitamin C is lost when canned peas are stored at room temperature, but 20% can be lost over a year when the cans are kept in a hot (80°F) store cupboard.

Dried Pulses—dried pulses do not regain all their water when cooked, and contain twice as much energy and slightly more protein, weight-for-weight than fresh pulses. Vitamin C, and most of vitamin A, are lost in drying, but thiamin is well retained. After soaking in water, dried (but not split) pulses can be sprouted, when vitamin C is formed in the shoots. On average, pea shoots contain 110 milligrams of vitamin C per 100 grams, mung beans (black gram) 100 milligrams, kidney beans 30 milligrams and soy beans 10 milligrams.

Baked beans are a variety of imported haricot beans dried before canning. They are as low as other vegetables in thiamin, but the tomato sauce supplies vitamins A and C.

PYRIDOXINE—or vitamin B$_6$—group name for

> pyridoxine (also called pyridoxol), pyridoxal, and pyridoxamine
> —obsolete names, adermin, factor I and factor Y

Part of the **vitamin B complex**, pyridoxine is essential for growth, blood formation, protection against infection, and healthy skin and nerves, but there is no adult disease specifically attributed to dietary lack of the vitamin.

All foods, except sugar, fats and spirits, contain small amounts of pyridoxine. Good sources are yeast (1 milligram per 25 grams); liver, kidney, mackerel (0·2 milligrams per 25 grams); meat (0·1–0·3 milligrams per 25 grams); fish

(0·05–1·5 milligrams per 25 grams); eggs (0·1 milligrams each); beer (0·2 milligrams per ½ pint) and milk (0·1 milligrams per ½ pint). Bananas, peanuts, pineapple and wholemeal bread also contain about 0·1 milligrams per 25 grams. However, about three-quarters of the vitamin is removed when white flour is milled—white bread only contains 0·02 milligrams per 25 grams.

Pyridoxine leaches out of foods cooked in water, like other B vitamins, and is also sensitive to heat, light and alkalies. There have been no extensive studies on the resistance of the vitamin to cooking and processing of food, but in general, acid foods (like fruits and some vegetables) retain the vitamin well. About one third is lost when meat is cooked, but there are no estimates of cooking losses in fish and vegetables. There are probably little or no losses in canning, but the vitamin may decline by as much as 30% during storage over a year in meat and vegetables. Canned orange, tomato and pineapple juices contain the same as fresh.

Between 1 and 1½ milligrams of pyridoxine are needed each day, but more is required (and eaten) when the diet contains more than 100 grams of protein. There are no British **recommended intakes** set for pyridoxine (due to lack of information) but the average daily intake of 1 to 2 milligrams is presumed to be sufficient for most people. More is required during pregnancy and breast feeding. Wholemeal bread is a convenient way of obtaining extra pyridoxine: 6 slices contain more than 0·5 milligrams.

Vitamin B_6, or pyridoxine, is the group name for three, equally potent, forms of the vitamin—pyridoxine, pyridoxal, and pyridoxamine. Pyridoxine is found mainly in cereals and vegetable foods, the others in animal foods. During **digestion**, nearly all the vitamin is transferred to the blood stream and transported to all cells of the body, where it is converted to its active form.

As part of a co**Enzyme**, pyridoxine has a vital role in the formation of new body **proteins** needed for growth and repair. It is needed to activate enzymes which recycle **amino** acids forming proteins (including haemoglobin—the oxygen carrying pigment in blood); regulatory substances in the brain (like dopamine, serotonin, or adrenalin); nicotinic acid; histamine (from histidine). It also takes part in the release of glucose from glycogen and possibly the conversion of linoleic acid into other essential fatty acids.

Pyridoxine is not stored, but it is contained in all cells and a diet totally lacking in the vitamin takes one to two months to deplete the body. Animals depleted of pyridoxine suffer a variety of disorders (including in monkeys a type of atherosclerosis and dental caries, which may be of relevance in human heart disease and tooth decay). It is possible that the diabetes which sometimes develops in pregnancy may occasionally be due to pyridoxine deficiency, but perhaps the only well established effect of a diet lacking in pyridoxine occurred when babies were fed a proprietary baby milk which had been subjected to severe heat treatment (oven drying). The babies suffered convulsions due to insufficient formation of GABA (a brain regulator). Fortunately the convulsions were cured by injections of pyridoxine.

Other adult deficiency symptoms have occurred when a diet lacking in pyridoxine has been augmented with an antagonistic drug. Additional symptoms included eczema around the eyes, mouth, scalp, neck and groin; a sore mouth; irritability, depression, drowsiness, nausea and sometimes neuritis. Similar symptoms of neuritis (and the 'burning feet syndrome'—see Pantothenic acid) often occur as a side effect of drugs used to treat tuberculosis.

Large doses of pyridoxine are harmless—the excess is filtered out of the blood stream by the kidneys. Ten to 100 milligrams daily are used in the treatment of two rare Inborn Errors of Metabolism: in adults a special type of anaemia can be cured, and in children (if the disorder is detected in time) epilepsy and mental retardation can be avoided. Large doses have also been used to treat radiation sickness, morning sickness in pregnancy, and nausea from anaesthetics. They may also be of benefit to some women who become depressed when taking the contraceptive pill, which sometimes interferes with the production of serotonin. However it is unlikely that severe depression, necessitating an admission to hospital, can be overcome by extra pyridoxine. The vitamin is equally ineffective in the treatment of seborrhoeic dermatitis.

RAFFINOSE

A sugar, one-quarter as sweet as **sucrose**, found in beet molasses. It is partially split to its constituent simple sugars (glucose, fructose, and galactose) when fermented by bacteria in the large bowel and may have to be avoided in **galactose**-free diets.

RAISING AGENTS

Bicarbonate of soda and an acid (cream of tartar, tartaric acid, acid calcium phosphate, acid sodium phosphate) added to cakes, scones and soda bread. When mixed with water or put into a hot oven, the gas carbon dioxide is released and expands, raising the mixture. In the baking industry raising agents are mixed just before baking. Baking powders are ready mixed and have **starch** added to separate the ingredients. They become less effective once the tin is opened. Golden baking powders contain added yellow colour (originally to simulate eggs) and release slightly less carbon dioxide than white ones. Self-raising flour also contains baking powder: it should be stored in a dry cool place.

Bicarbonate of soda destroys **thiamin** (vitamin B_1): the greater the amount used the greater the destruction. Soda bread (and some bread mixes), cakes and scones are very unreliable sources of vitamin B_1. Where possible, bread should be leavened with yeast.

RECOMMENDED INTAKE—or allowance

The quantity of a **nutrient** in the average daily **diet** which will maintain health (prevent deficiency disease) in practically all healthy people in a given population. Tables of recommended intakes are set for different groups of people in the population, according to age, sex, and occupation. There are special provisions for pregnancy and breast feeding. The British recommended intakes, set by the Department of Health and Social Security, are shown in the Appendix. Each individual recommended intake assumes that all other recommended intakes are met, and does not cover increased needs caused by underlying disease (like some **Inborn Errors of Metabolism**).

Recommended intakes for all nutrients, except **energy**, are greater than the average requirement (the least amount that will maintain health) for the group. Even when age, sex, and occupation are taken into account, there is great natural variation between individuals. To allow for this variation, and for the availability of nutrients from different foods, recommended intakes include 'safety factors' added on to the average requirement. Thus, for many people, the recommended intake will be greater than actual needs: an individual eating less than the recommended intake will incur a greater risk of ill health, but will not necessarily be unhealthy. However, the recommended intakes—since they are designed to cover everyone in a given population—are unlikely to be too low for an individual. Consequently supplements, as pills and tonics, taken over and above a **balanced diet** are unlikely to be of benefit unless there is some underlying disease.

In the case of **protein,** the recommended intake is even greater than the average requirement plus 'safety factors'. For instance, for an adult man, the average requirement can be calculated fairly accurately. It is about 30 grams per day. With 'safety factors', 20% for natural variation and 14% to allow for the fact that protein in the average British diet is

not used with 100% efficiency, it is about 45 grams per day.
However, in Britain, the recommended intake depends on an
arbitrary ratio of protein to energy. It is 68 grams for a man
living a sedentary life. See Protein for more information.

In the case of energy, no 'safety factors' are included, and
the calculated average requirement of a group of people is
taken as the recommeneded intake. Clearly, because of
natural variation around the average, many people in the
group will need more than the recommended intake;
others will need less. See Energy for more information.

There is lack of research material on which to base a
recommended intake for some nutrients—usually because
the minute quantities required are found in most foods and
human deficiency caused by diet is unknown or extremely
rare. If the recommended intake for the main nutrients are
met, then the diet is likely to contain sufficient of these other
nutrients. However, when the diet contains many new foods
composed of refined materials, this supposition may not be
true. Reputable manufacturers may add the main nutrients
lost in processing, but little attention is paid to the lesser
known nutrients. Some fresh unprocessed food should always be included in the daily diet. Other comparatively
recent developments—like new breeds of animals and plants
and possible changes in soil composition—may also alter
nutrients in food. A wide variety of foods eaten in moderation minimises any possibility of dietary deficiencies.

RIBOFLAVIN—other name vitamin B_2
                       —obsolete names vitamin G and lactoflavin

Riboflavin is part of the **vitamin B complex** and needed for
growth in children and to maintain a healthy skin and eyes.
It is present in nearly all foods except sugar, fats and spirits.

Liver, kidney, heart and **extracts** are rich sources of riboflavin: cheese, eggs, milk, mackerel, roes, mushrooms,
'enriched' breakfast cereals, and chocolate are good sources.
Wholemeal bread, meat, fish, peanuts, dark green vegetables,

prunes, beer and avocado pears contain moderate amounts: white bread, unenriched breakfast cereals, fruits and other vegetables are poor. In the average diet, 40% of the total intake of 1·79 milligrams of riboflavin is obtained from milk and cheese, and 20% from meat.

In cooked and processed foods, riboflavin is fairly stable to heat, but it is destroyed by ultra-violet light, is sensitive to alkalies, and, like other B vitamins, leaches out of foods cooked in water.

The most drastic losses of riboflavin occur when milk is exposed to light. Half the vitamin is lost in two hours' exposure to bright sunlight, and one-fifth in dull light. Milk on sale in clear containers under fluorescent lighting is likely to contain none. Slight losses occur when eggs and milk are cooked in an open pan.

Some riboflavin is lost from meat and fish during cooking, but nearly all can be recovered if the liquor is used for sauce or gravy. Canned and frozen meats retain most of the vitamin: there are no losses in stored canned meats, but the vitamin may decline (by up to 30% in 8 months) in frozen meats. There may also be slight losses in the drip of frozen meats. Riboflavin also leaches out of vegetables but provided bicarbonate of soda is not used (bicarbonate is an alkali) the vitamin is also recovered in the cooking water. Bicarbonate of soda in baking powder may also destroy some riboflavin in soda bread and cakes.

Needs for riboflavin are related to the number of active cells (all except fat) in the body. Athletes and other people with highly developed muscles will need more than others. The minimum daily needs for a man are about 1·1 milligrams, but to allow for individual variation, the daily **recommended intake** for men is 1·7 milligrams. The recommended intake for women (who have less muscle than men) is less—1·3 milligrams. The table overleaf shows portions of foods supplying this amount. Children, infants and adolescents need more in proportion to their weight than adults to allow for growth, and women need more than usual during pregnancy and breast feeding to ensure adequate supplies for the child.

Source	Food	Average content milligrams per 100 grams	Quantity supplying 1·3 milligrams	Approximate measure
Rich	Liver	3	45 grams	1½ oz
	Kidney	2	65 grams	2½ oz
Good	Cheddar cheese	0·5	260 grams	½ lb
	Egg	0·35	370 grams	6 large
	Milk	0·15	870 mls	1½ pints
Moderate	Roast beef	0·22	600 grams	1¼ lb
	Boiled broccoli tops	0·2	650 grams	1¼ lb

Nearly all riboflavin is absorbed out of food during digestion and transferred to the blood stream. As part of two very important coEnzymes in cells, riboflavin is essential for many body processes, particularly the release of energy from food. Lack of the vitamin checks growth in children, but surprisingly does not appear to have very serious effects in adults. In adults the main symptoms are a sore magenta tongue; a type of eczema round the nose, chin and groin, and painful fissures (angular stomatitis) around the mouth. The eyes may become bloodshot and itchy. None of these symptoms is specific for riboflavin deficiency—for instance, angular stomatitis can be caused by poorly fitting dentures.

Large intakes of riboflavin are not harmful—the excess is filtered out of the blood stream by the kidneys—but are inadvisable. If the diet is inadequate with respect to riboflavin, it is also likely to be lacking in other nutrients. An improvement in diet (more liver, dairy products and green vegetables) are preferable to vitamin pills and tonics. The elderly are perhaps the most vulnerable to riboflavin deficiency and should include at least ½ pint of milk in their daily diet.

RICE

Like other cereals, rice contributes energy, protein and most B vitamins to the diet, but cannot support life when eaten alone because it is lacking in vitamins A, C and B_{12}. It contains rather less protein than other cereals, and no **gluten**.

Brown rice has only the outer husk removed and contains all the nutrients in the germ and outer layers of the grain. White rice is a very poor source of B vitamins: for instance 80% of the thiamin is removed. Glossy polished rice is pro-

AVERAGE NUTRIENTS IN PORTIONS OF SOME RICE PRODUCTS

Nutrient	in 60* grams (2 oz) uncooked white rice	in 60* grams uncooked brown rice	in 25 grams puffed rice
Energy (kilojoules)	905	905	370
(kilocalories)	215	215	85
Protein (grams)	4	4½	1½
Carbohydrate (grams)	48	46	21
Fat (grams)	¼	1	¼
Sodium (milligrams)	5	5	200a
Potassium (milligrams)	70	90	35
Calcium (milligrams)	15	20	1
Magnesium (milligrams)	15	90	10
Phosphorus (milligrams)	55	135	30
Iron (milligrams)	0·5	1	0·2
Other minerals	Small amounts – brown rice more than others		
Vitamins A, C, D	0	0	0
Vitamin E (milligrams α tocopherol)	0·05	0·2	—
B vitamins			
thiamin (milligrams)	0·04	0·17	0·3a
riboflavin (milligrams)	0·02	0·03	0·35a
nicotinic acid (milligram equivalents)	1	1	4a
pyridoxine (milligrams)	0·2	1	—
pantothenic acid (milligrams)	0·4	0·8	—
folic acid (micrograms)	10	—	—
biotin (micrograms)	1·8	—	—
vitamin B_{12}	0	0	0

— – no figures available
a – added in processing
* – enough for one person. About 15 grams are used in a milk pudding

duced by a further rubbing in leather-lined drums—sometimes with talc and glucose. Parboiling before milling and polishing, when the B vitamins diffuse into the centre of the grain, causes less loss of thiamin—between 30 and 60%, depending on the extent of polishing.

Most people in Britain eat so little rice that losses in milling do not affect the quality of the diet as whole. It is normally eaten—as a breakfast cereal or pudding—with milk, a good source of essential nutrients. Puffed rice is low in B vitamins, but some manufacturers add thiamin, riboflavin and nicotinic acid.

When rice is used as an alternative to potatoes, brown rice is preferable. Weight-for-weight, boiled brown rice provides about as much thiamin as boiled potatoes (and slightly more energy).

RICKETS

Usually caused by lack of **vitamin D**, normally made in the skin after exposure to ultra-violet or sunlight. Most food supplies too little vitamin D for children's needs—for instance, a child would have to drink 35 pints of milk or eat $1\frac{3}{4}$ lb butter or 11 eggs or $4\frac{1}{2}$ oz margarine to fulfil the daily **recommended intake**. Fatty fish are the only good sources (4 oz supplies a child's recommended intake).

When there is lack of vitamin D, insufficient **calcium** is absorbed from food and the growing child's bones are not hardened and become deformed. The skull is 'bossed' instead of round; there may be a pigeon chest and a line of bumps on the ribs; and the wrist, knee and ankle joints are enlarged. When the child begins to walk, the long bones in the legs bend (resulting in bow legs), and the backbone and pelvis are twisted. The deformities in the chest and pelvis carry a greater susceptibility to bronchitis, and for girls, later difficulties in bearing children.

Rickets is thought to have been recognised in the sixteenth century, and was well known in the seventeenth century. It was particularly common in the darkest days of the Industrial Revolution, and, at the beginning of the twentieth century 75% of children living in cities are said to have been affected. After the cause was elucidated, clean air acts, better housing, fortification of margarine and baby milks, and distribution of welfare cod liver oil almost eliminated the disease.

However, rickets is still a problem in Britain (Asians and children confined to high rise flats are particularly at risk), occurring most frequently in toddlers. Some children, whose mothers have **osteomalacia**, are even born with it. All children require a daily supplement of vitamin D (as welfare vitamin drops or cod liver oil) up to the age of five. Failure to give the supplements, coupled with inadequate exposure to sunshine is thought to be the explanation of the continuing (and increasing) problem. West Indian and other coloured immigrants are apparently less affected than Asians, probably because they tend to wear Western dress (which exposes more skin) and possibly make greater use of margarine. Schoolchildren and adolescents are also affected: any child who has little sunshine will benefit from extra vitamin D (as margarine, cod liver oil or malt) until growth is completed. Excessive doses are toxic.

Other types of rickets, caused by inherited abnormalities, and kidney disorders, are not responsive to the usual dose of vitamin D.

ROUGHAGE
See Fibre.

RYE

A **cereal,** with similar nutritive properties to wheat, that will grow in cold climates and is an important crop in Scandinavia, Russia and north Germany. It contains a small quantity of **gluten,** and rye flour can be made into bread. Rye bread has a sour taste from **lactic acid** (made by bacteria in the yeast mixture) and sometimes added **citric acid**. Pumpernickel and black breads are made entirely from rye, but others contain 40 to 80% wheat flour. Nutrient contents depend on the proportion of wheat, and the extraction rates

AVERAGE NUTRIENTS IN 25 GRAMS (1 OZ) RYE CRISPBREAD
AND RYE FLOURS

Nutrient	Crispbread	Dark rye flour	Light rye flour
Energy (kilojoules)	360	350	370
(kilocalories)	85	85	90
Protein (grams)	1¾	2	1½
Carbohydrate (grams)	20	20	22
Fat (grams)	½	½	¾
Sodium (milligrams)	155a	—	—
Potassium (milligrams)	115	105	35
Calcium (milligrams)	10	10	5
Magnesium (milligrams)	25	25	5
Phosphorus (milligrams)	75b	90b	20
Iron (milligrams)	0·9	0·7	0·3
Vitamins A, C, D, B₁₂	0	0	0
Vitamin E	—	—	—
B vitamins			
thiamin (milligrams)	0·09	0·1	0·04
riboflavin (milligrams)	0·06	0·05	0·02
nicotinic acid (milligram equivalents)	0·3	0·3	0·2
pyridoxine (milligrams)	—	0·09	0·02
pantothetic acid (milligrams)	—	0·3	—
biotin (micrograms)	—	1·5	—
folic acid	—	—	—
Other minerals	Small amounts		

— – no figures available; a – salt added in processing; b – mostly as phytic acid
which interferes with the absorption of iron, magnesium and calcium

of the **flours** used. Coarse (heavy dark) rye flour is similar to
wholemeal flour, light, similar to white unenriched wheat
flour. Rye bread may be coloured with caramel, and darker
loaves are not necessarily higher in nutritive value.

In Britain, most rye is eaten as crispbreads, which are very
dry and consequently 'light', but weight-for-weight contain
more **energy** than wholemeal bread. In slimming diets, no
more than 2 medium or 3 thin crispbreads should be exchanged for each slice of bread.

SAGO

Extracted from the pith of the sago palm and sold as 'pearls'.
Its poor nutritional value—it is virtually pure **starch**, yielding
energy but little other nutrients—is improved when cooked
with milk.

AVERAGE NUTRIENTS IN 15 GRAMS (½ OZ) SAGO

Energy (kilojoules)	225
(kilocalories)	55
Carbohydrate (grams)	14
Protein (grams)	0·03
Fat (grams)	0·03
Iron (milligrams)	0·2
B vitamins (nicotinic acid) (milligrams)	0·03
Other vitamins and minerals	0 or trace only

SALT—chemical name sodium chloride

The main source of sodium in the diet, it is used as a **preservative** and flavour enhancer in food.

A normal diet, containing salty foods and salt in cooking supplies about 10 grams of salt. Table salt, added after food is cooked, may contribute another 4 to 10 grams per day. A high sodium diet is associated with high blood pressure. Except in very hot climates, it is worthwhile to avoid excessive use of table salt.

Free-running table salt contains up to 2% of anticaking agents, like magnesium carbonate. Iodised salt should contain between 433 and 725 micrograms of **iodine** per ounce: a half level teaspoon (3 grams) iodised salt supplies 45 to 75 micrograms iodine—enough to fulfil half the assumed daily needs. Unless iodised, sea salt contains only small quantities of iodine, in an unstable form which disappears on storage.

Salt substitutes contain **potassium** salts in place of sodium chloride, and can be used in some low sodium diets. They must be avoided in low potassium diets—used for some types of kidney diseases.

SCURVY

The result of a diet totally lacking in **vitamin C**, necessary for the integrity of collagen, the main **protein** in connective tissue which 'holds' the body together.

Tiredness and depression precede the characteristic features of scurvy—haemorrhages—which are first noticed as

bruises on the skin. The gums swell, ulcerate and bleed, and
the teeth loosen and fall out. Old wounds, which are knitted
together with connective (scar) tissue, reopen and the legs
become blue and swollen. Death follows rapidly from internal haemorrhage unless vitamin C is given, when there is a
dramatic improvement in health and vigour.

In Britain, scurvy is rare but occasionally occurs in
babies. Bottled, evaporated and some dried milks contain
virtually no vitamin C and it must be given as welfare vitamin drops or orange juice (not boiled). Breast milk may be
an inadequate source if the mother's diet is low in vitamin
C. In adults, any diet lacking in fruit or vegetables (like the
'higher' levels of the Zen macrobiotic diet) is dangerous.
Scurvy has also occasionally resulted from other very restricted diets, for instance milk regimes for ulcers. Many
elderly people—subsisting on meals containing overcooked
vegetables and fruits poor in vitamin C may be on the bor-
derline of scurvy. Instant mashed potatoes should be avoided
unless they have added vitamin C (declared on the label).

Scurvy was common in Medieval North Europe (particularly in early spring) but declined after the introduction
of the potato. However, as late as the early twentieth-cen-
tury military campaigns, sea voyages and expeditions were
all liable to be jeopardised by scurvy. Florence Nightingale
believed that more men died from it in the Crimea than
from any other cause, and at least one member of Scott's
last journey to Antarctica is thought to have died from
scurvy. Not until the discovery of the vitamin in the 1920s
was it fully realised that the disease was a result of deficiency
in food, rather than infection, salt, poisons, or the many
other factors associated with it.

SELENIUM

A rare **trace element**, known to be an essential nutrient for
animals, and thought to be essential for humans. Like other
trace elements, selenium is toxic in excess.

Soil contents of selenium—which vary markedly in different parts of the world—determine the quantity eaten in food (plants take up selenium from soil). Water contents, which reflect soil contents, vary from 1 to 300 parts per million. The amount necessary for human health is unknown, but British diets are assumed to be adequate: levels of selenium in blood samples from British people were found to be three times higher than in Sweden, a low selenium area.

Interest in selenium was aroused when it was discovered that it could prevent some of the symptoms of **vitamin E** deficiency in animals. More recent experiments have demonstrated that selenium is an essential element in its own right for all animals studied. Monkeys fed a diet lacking in selenium, but adequate in vitamin E, lost weight and hair, became listless and died from liver, kidney and heart failure. Farm animals feeding on low selenium pastures develop a type of muscular dystrophy, cured by adding amounts to the feed. It is likely that it plays a fundamental role in the maintenance of the structure of cells but no specific disease in humans has so far been attributed to lack of selenium. It has been suggested that a very low intake predisposes to cancer but this claim has been criticised for its lack of scientific evidence.

Pastures of high selenium areas may be toxic to animals (causing a disease called blind staggers) but, apart from a possible increase in dental decay, no disease caused by excess selenium has been attributed to humans.

SILICON

The second most abundant element: one-quarter of the earth's crust is silicon. Traces are found in the human body, but it is not known if it is an essential nutrient. Recent reports have suggested that it is for some animals.

Some salts of silicon are permitted in food as miscellaneous **additives**, for example calcium silicate, aluminium silicate (kaolin) and magnesium trisilicate are permitted anticaking agents. Dimethylpolysiloxane—an organic (carbon containing) polymer of silicon—is a permitted antifoaming agent: it prevents unwanted foams when, for example, fruit juices are canned.

SLIMMING DIETS—see also Obesity

The principle behind *all* slimming regimes is the same: there are no slimming foods. If a diet contains less than daily needs, **energy** is withdrawn from stores (mostly fat) in the body and weight decreases. Each kilogram (2·2 lb) of human fatty tissue contains about 7000 kilo**calories**. If a slimming diet contains, for instance, 1000 kilocalories less than daily needs, over the week there is an energy deficit of 7000 kilocalories. The weekly expected weight loss (the amount of fat used from stores) will be 7000/7000—which is equal to 1 kilogram.

Anyone can reduce their weight simply by eating less, but the energy and nutrient values of foods are not easily remembered. Most people need to work within a definite plan. Examples of two low calorie diets, which will not deplete the body of essential nutrients, are shown in the table, pages 302–3. Vitamin tablets are not necessary. Any reasonably active person will lose weight on the 1500 kilocalorie diet, but the rate of loss may be too slow for some women and small men. It is important however not to limit energy in the diet to below 1200 kilocalories when first starting to slim: some people appear to adapt to low energy levels.

Low calorie diets allow unlimited amounts of leafy vegetables and some drinks (see 'free' list) which contain very few kilocalories. Moderate energy containing foods—like meat, fish, cheese, eggs, potatoes, bread and milk and fruits—which supply essential nutrients, are allowed in restricted amounts. They *must* be weighed out at the beginning until their weights can be accurately judged by eye. Overestimations are very easily made, especially when hungry. A small scale (no. 18D, made by Salter Ltd) intended for use by diabetics can be bought from chemists. Fats (except for a small allowance of butter or margarine) and sugar (and foods that contain them—see 'foods to be avoided' list) are excluded. They are high in energy but supply little or no essential nutrients. Saccharin (but not other **sweeteners**—like **fructose, sorbitol**) can be used for sweetening fruit—both liquid and solid saccharin is available. To avoid bitterness, saccharin should be added after cooking. Low Calorie carbonated drinks (like Slimline) with ice and lemon are socially inconspicuous alternatives to alcohol.

At the outset of dieting, stocks of the 'foods to be avoided' must be thrown away. It is also better to avoid shopping for food when hungry. Eating food with the greatest possible concentration and with the slowest possible speed helps the slimmer recognise fullness: any food left on the plate should be discarded. Drinking water or other low Calorie drinks with meals also helps to promote satiety.

When eating out, choices from the free list should be made for starters (for example melon, fresh grapefruit, tomato juice). Portions of meat for the main course must be very carefully judged (a large steak can contain 500 kilocalories which is a half to a third of the total allowance for the day) and any thick or creamy sauce must be left. Boiled potatoes and salad or vegetables can be chosen with the main dish. If no dessert fruit is available, fruit salad (without cream) or ice cream contain less kilocalories than cheese and biscuits.

A check on weight, wearing the same clothes and at the same time of day should be made once a week, and recorded on a chart. Many people lose weight very quickly in the first few days of dieting. This is fluid, released when **glycogen** is used up to compensate for less carbohydrate in the diet. Very low carbohydrate diets (see Ketogenic diets page 305), promising quick results, are based on this effect. Fat is used for energy once glycogen has been lost, but it takes twice as long to lose each pound of fat as it does to lose each pound

LOW CALORIE SLIMMING DIETS

	For both diets		For 1500 kilocalories		In addition — For 1200 kilocalories	
Meal	Portion of food	Approx. kcals.	For 1500 kilocalories	Approx. kcals.	For 1200 kilocalories	Approx. kcals.
Breakfast	Fresh grapefruit or unsweetened grapefruit juice if wanted	20	Bread, 2 large slices (2½ oz), toasted if wanted, preferably wholemeal	200	Bread 1 large, slice (1½ oz) toasted if wanted after weighing, preferably wholemeal	100
	Egg, scrambled, poached or boiled—one or lean grilled bacon or ham—1 oz or kipper—2 oz or smoked haddock—3 oz	70				
	Tomatoes or mushrooms, tinned or grilled	10				
	Butter or margarine from allowance					
Lunch	Starter from 'free' list if wanted		Potato, boiled 4 oz or bread 1½ oz or rice, boiled 3 oz or pasta, boiled 3 oz or crispbread 1 oz	100	Potato, boiled 2 oz or bread 1 oz or rice, boiled 1½ oz or pasta, boiled 1½ oz or crispbread ½ oz	50
	Lean meat—3 oz or white fish—6 oz or eggs—2 or cheese—1½ oz (or 6 oz cottage) or 3 oz fatty or tinned (no oil) fish	180				
	Large helping vegetables from 'free' list	10–30				
	Large helping peas, beans or sweetcorn					
	1 medium apple or orange or pear (4 oz) or 1 small banana (2 oz) or grapes, tinned fruit drained of syrup—2 oz	40				
Evening Meal	As lunch	230–250		100		50

Meal	For both diets		In addition			
	Portion of food	Approx. kcals.	For 1500 kilocalories	Approx. kcals.	For 1200 kilocalories	Approx. kcals.
Allowances	Butter or margarine 1 oz or low fat spread 2 oz	220	Milk, fresh ¾ pint or skimmed milk 1½ pints or yoghurt 1 pint (plain) or dried skim milk 3 oz	300	Milk fresh ½ pint or skimmed milk 1 pint or yoghurt ½ pint (plain) or dried skim milk 2 oz	200
Drinks during the day	As wanted from 'free' list	770–810		700		400
			Total 1470–1510		1170–1210	

Food may be cooked with butter and milk from allowance, but with no thick gravy or sauce. Allowed freely—tea and coffee (black or with milk from allowance) Bovril, Marmite, Oxo, vinegar, water, salt, pepper, spices, saccharin, gelatine, clear soup, tomato juice, soda water, diabetic squash, sugar free carbonated drinks, all leafy vegetables, carrots, cauliflower, celeriac, celery, cucumber, mushrooms, onions, all herbs (parsley etc), pumpkin, radishes, swedes, tomatoes, turnips, salsify, and, without sugar, cranberries, redcurrants, lemons, melon, rhubarb, fresh grapefruit, lemon juice (saccharin may be used instead of sugar for sweetening fruits).

Foods to be avoided—Sugar, glucose, confectionery, jam, honey, marmalade, syrup, dried or crystallised fruit, dried pulses, baked beans, puddings, cakes, pastries, biscuits and cereals, fried foods, fatty meat, sausages, cream, cream cheese, beer, wines, spirits, soft drinks, fruit yoghurt, thick soups, nuts, cocoa, Ovaltine, Horlicks, drinking chocolate, diabetic or 'slimming' foods, except for diabetic squash.

of glycogen (fat contains twice as much energy as glycogen). There is also little water lost from fat cells and the apparent halt in weight after the first few weeks is disheartening. Many people give up at this stage.

With determination however, most people can lose their fat. It will take the 85 kilogram (13 stone) man about 2 months to lose 6½ kilograms (one stone) on a 1500 kilocalorie diet. If weight loss halts for more than 3 weeks, the energy in the diet should be reduced by about 100 kilocalories (for example by cutting out a slice of bread) at a time. A diet very low in energy (less than 800 kilocalories a day) should not be taken without medical advice. When this very low level is reached, exercise must be increased to continue weight loss. A brisk walk for half an hour will use up 150 kilocalories.

Once ideal weight (see Obesity) has been reached, great care is needed to prevent weight returning. It is best to replace energy slowly in the diet (in weekly steps of 200 kilocalories) up to 2000 kilocalories for women and 1500 kilocalories a day for men. These intakes are slightly lower than the recommended intakes for people leading sedentary lives. If weight continues to be lost the intakes should of course be increased.

Long-term stability can only be achieved by a change in attitude to food. It becomes more difficult to rationalise dietary indiscretions—like overeating and nibbling—once the energy values and nutritional facts of foods are thoroughly learned. See Energy, Obesity, Balanced diet, and individual food entries.

Other slimming regimes

Some diets attempt to sugar the pill by allowing the slimmer to eat unlimited amounts of selected foods. Others restrict choice to a very narrow range.

Low carbohydrate diets allow unlimited quantities of meat, fish, cheese, eggs, butter, cream and other fats and oils. These foods contain no **carbohydrate**. Leafy vegetables and certain drinks (see 'free list' in low Calorie diets) are also allowed as wanted. All concentrated sources of carbohydrate (sugar, starches and foods containing them—cakes, biscuits, sauces, pastry, confectionery, preserves) are excluded. Other sources—milk, fruit, potatoes, cereals, bread—are allowed in very restricted amounts. Most regimes limit the carbohydrate to 50 to 80 grams—a typical diet is shown opposite. Using the exchange system (portions of foods containing 10 grams of carbohydrate—see list in carbohydrate section) it is easy to vary the diet. For instance, a small portion of breakfast cereal and a glass of milk (but no sugar) could be taken in place of bread for breakfast. Other low carbohydrate diets—like *This Slimming Business* by John Yudkin (Penguin) and some American books—use 5 gram exchange units: portions of foods containing 5 grams of carbohydrate.

By reducing carbohydrate, the total energy in the diet should, in theory, be reduced. For example the average diet contains 300 grams of carbohydrate. In the sample diet, only 80 grams are permitted and the reduction in energy from the average will be $4 \times (300 - 80)$ which equals nearly 900 kilocalories (each gram of carbohydrate would have supplied 4 kilocalories). Indirect restriction of fat (in cakes, biscuits etc) also lowers the energy intake. However, it is possible to *gain* weight with a low carbohydrate diet: cheese, meat and fats are high in energy and very large portions overcompensate for the reduction with less carbohydrates.

Ketogenic diets. For example Dr Atkins Diet Revolution, the Drinking Man's Diet, the Air Force Diet, the Quick Weight Loss Diet, the Grapefruit Diet and some so called 'Mayo Clinic' Diets. All are very low carbohydrate regimes (0 to 20 grams per day) unrestricted in protein and sometimes fat. They force the body to release large quantities of fatty acids (from fat stores—see Lipid), which can only be partly used for energy. The unused part (ketones) accumulate in the blood stream, causing nausea. Appetite is reduced

LOW CARBOHYDRATE DIET

Meal	Food	Carbohydrate grams	Exchanges Numbers
Breakfast	Eggs, bacon, as wanted, with fried, tomatoes or mushrooms	—	—
	One orange or 4 oz unsweetened orange juice	10	1
	Bread, one large slice	20	2
	Butter, margarine as wanted	—	—
	Black tea or coffee, no sugar, with milk from allowance	—	—
Drinks in day	As for breakfast, or any of those permitted on free list on low Calorie diet	—	—
Lunch	Cheese, egg, fish, meat, as wanted, with butter, fried, or with cream, but no starch or flour	—	—
	Vegetables from free list with butter if wanted	—	—
	One apple, or two cream crackers and cheese	10	1
Supper	Main course as lunch	—	—
	Vegetables or salad from free list with butter or vinegrette	—	—
	2 tablespoons peas or beans	5	½
	4 oz fresh fruit salad with cream	10	1
Before Bed	1 glass milk	10	1
	1 plain biscuit	5	½
Allowance of milk for drinks	½ pint	10	1
Total		80	8

and less food is eaten. Some of these diets include strict
instructions to eat particular combinations of foods, claimed
to hasten fat loss. It should be remembered however that
there are no known slimming combinations of food: all the
diets probably rely on this basic action. Any diet containing
too little carbohydrate will cause excessive fatty acid release:
diets which encourage plenty of fat on food (like the Atkins
regime) make ketone production more likely. The 'Diet
Revolution' has been severely criticised by the American
Medical Association: **gout** may be exacerbated, and blood
lipids may rise, increasing the risk of a heart attack. Fluid
(released as glycogen is used) accounts for weight losses of
about 4 kilograms claimed for '10-day' diets—for example

the grapefruit diet. As soon as former eating habits are resumed after the diet, glycogen and water reaccumulate and weight returns.

Other diets. Crash diets based on a rigid choice of a few foods (for example the banana and milk diet, formula drink diets) are low in energy but—since any successful diet must be taken for a long time—carry the risk of a deficiency of essential nutrients. However, most people cannot tolerate severe curtailment of choice for long and eventually return to former habits, regaining any weight lost almost immediately.

Slimmers meals, puddings, soups, sweets and breads are often no lower in energy than normal foods, and are of limited value in controlling weight over the long term. It is better to establish good eating plans with normal food, rather than perpetuate fattening habits.

Acids (grapefruit, lemon juice, vinegar) are sometimes claimed to 'burn up fat'. This is a fallacy. If these substances do help to reduce weight it is due to their effect of aiding some people to overcome a sweet tooth. Some diets prohibit caffeine (in coffee), others recommend it—a reflection of caffeine's lack of any marked effect in slimming diets. Other diets restrict **salt** and fluid, causing a loss of weight, but no fat. Except under medical advice, there is no need to exclude salt, and water should never be restricted.

SODIUM

An **electrolyte,** sodium is of vital importance in maintaining a constant body **water** content. Most sodium in the diet is derived from **salt** (sodium chloride).

Some foods—meat, fish, eggs, and milk—naturally contain small amounts of sodium. Shellfish contain slightly more than ordinary fish. A daily diet to which *no* salt is added supplies about 0·5 grams of sodium. Flour, unsalted butter, cream, oil, fruits, fresh vegetables, and unsalted nuts contain virtually none.

Salt, added in cooking, at the table, and to many foods, causes wide variation in the sodium content of individual diets. Most people eat between 4 and 8 grams of sodium a day. The table below shows portions of salted foods supplying approximately 0·5 grams of sodium. Other sources of sodium are **monosodium glutamate** and bicarbonate of soda in mineral waters and baking powder.

Foods to which salt is added include all cured meats and fish; sausages; pickles; sauces; canned and bottled meats; vegetables and fish; most breakfast cereals; crisps; olives; roasted nuts; bread; salted butter; margarine; shellfish; extracts; packet of soups; tinned soups.

Adults contain between 75 and 100 grams of sodium. Over half is in the blood plasma and fluids surrounding cells (interstitial fluids); 10% is held inside cells; and the rest forms part of the structure of bones. The body content is kept constant: in temperate climates the quantity of sodium gained from food is balanced by equivalent losses in the urine.

This balance can be upset by excessive losses of salt in perspiration. Although the kidneys reduce the quantity lost in urine to almost zero, sodium lost in sweat can exceed the day's intake. In Britain sodium deficiency is only likely in people who work in hot conditions—like miners and steel workers—but mild symptoms, muscular cramps, can follow a few hours' exercise on a hot summer's day.

In very hot climates, severe deficiency (from a loss of 8 grams or more) has profound effects on the body. Water and sodium are tran ferred out of the blood stream (to maintain the concentration of sodium in interstitial fluid) when blood pressure falls, causing tiredness, nausea and dizziness. If thirst is satisfied with water (or other salt-free drink like beer) the body fluids are diluted, resulting in convulsions and coma. Salt should always be taken together with water after heavy perspiration.

Other causes of sodium deficiency are very low salt diets coupled with sauna baths (like some slimming regimes);

	Food	Approximate portion supplying 0·5 grams of sodium
Meat	Fried bacon	15 grams (1 rasher)
	Corned beef	35 grams (1¼ oz)
	Fried pork sausages	50 grams (1¾ oz)
Fish	Prawns boiled in salt water	30 grams (1 oz)
	Tinned sardines	60 grams (2 oz)
Dairy foods	Cheddar cheese	85 grams (3 oz)
	Salted butter	85 grams (3 oz)
Cereals	Cornflakes	50 grams (1¾ oz)
	Bread	100 grams (3 slices)
Miscellaneous	Table salt	1·3 grams (small pinch)
	Marmite	10 grams (⅓ oz)
	Tomato ketchup	50 grams

diarrhoea and vomiting; and certain hormonal and kidney disorders.

In adulthood, provided sufficient water is available, excess sodium is eliminated from the body by the kidneys. Sodium poisoning only occurs when sea water is drunk in the absence of fresh water. However, except when there is a possibility of deficiency, it is inadvisable to salt food heavily: diets high in salt have been shown to cause high blood pressure in animals and are linked with high blood pressure in humans. Highly salted foods and **extracts** should never be given to babies. Their immature kidneys require more water to eliminate excess sodium than adult kidneys.

Low sodium diets are part of the treatment of some heart and kidney disorders. In severe heart failure, blood is not pumped around the body with sufficient force, and the kidneys become less effective. Sodium accumulates and the body becomes waterlogged (oedematous). Diuretics (which increase the flow of urine) and a low sodium diet may be necessary.

All low sodium diets exclude table salt (salt added after food is cooked) and very salty foods like cured and pickled meats and fish; bottled sauces; cheese; sausages; bacon; ham; tinned meats and fish; salted nuts and extracts, but usually allow a small amount of salt in cooking. Stricter diets may exclude certain breakfast cereals (like cornflakes, Rice Krispies); cakes made with bicarbonate of soda or baking powder; and salted butter. Further restriction—no salt in cooking, salt-free bread, special milk—is avoided if possible because at this level the diet becomes very unpalatable.

SOFT DRINKS

Squashes, cordials, fruit drinks and carbonated drinks are of poor nutritive value: their main ingredients are **water** and **sugar**. Soft drinks are unlikely to contain **vitamin C**, unless it is added and declared on the label. For fruit juices see Fruit.

Legally, all soft drinks (except diabetic and low calorie drinks) must contain specified minimum contents of **sucrose** (sugar). Saccharin is permitted, up to specified amounts, but must be declared on the label. There are also specified minimum contents of fruit: those that contain no fruit (and are essentially flavoured and coloured water) must be called . . . flavour or . . . ade, and no picture of fruit is allowed on the label. Comminuted fruit drinks (usually described as made from whole fresh . . .) made from finely macerated pulp and peel, are allowed to contain less fruit than squashes, crushes and cordials. Citrus fruit and barley drinks like lemon and barley water may also contain less fruit than squashes.

Carbonated drinks have to comply with similar standards. Tonic water must contain a minimum quantity of quinine (which is usually the bitter ingredient in bitter orange and lemon). Glucose drinks like Lucozade contain glucose syrup in place of sugar. Cola drinks contain sugar, caramel (to colour brown), phosphoric acid, flavouring oils, carbon dioxide and usually **caffeine**.

All soft drinks are permitted to contain **preservatives** (either sulphur dioxide or benzoic acid), **emulsifiers** (to prevent particles of fruit clumping together and forming sediment), **colours, flavours** and miscellaneous **additives**. Only citric, malic, tartaric, ascorbic and nicotinic acids (added to give a tart flavour) are permitted in fruit drinks. Non-fruit drinks may also contain acetic and phosphoric acids.

'Diabetic' soft drinks must not contain sugar: they are

AVERAGE NUTRIENTS IN SOME SOFT DRINKS

Nutrient	in 60 grams fruit drink (undiluted)		in 200 grams (1 glass) carbonated drink		
	Orange/ Lemon	Lemon Barley	Tonic water	Bitter lemon/ orangeade	Glucose drink
Energy (kilojoules)	250	265	195	295	575
(kilocalories)	60	65	45	70	135
Protein, fat	0	0	0	0	0
Carbohydrate (grams)	16	17	12	19	36
Potassium (milligrams)	10	10	1	10	2
Sodium (milligrams)	15	20	10	15	55
Other minerals	Small amounts		Small amounts		
Vitamin C	Usually 0	0	0	0	0
Other vitamins	0	0	0	0	0

sweetened with saccharin and are virtually free of kilocalories. Low calorie drinks must legally contain not more than $7\frac{1}{2}$ kilocalories per fluid ounce (or $1\frac{1}{2}$ kilocalories per fluid ounce when sold ready diluted).

SOLVENTS—see Additives

Used for the incorporation of **flavours, colours** and other food **additives** into food. Nine are permitted by the Solvents in Food Regulations 1967:

Ethanol (alcohol)
Ethyl acetate (combination of alcohol and acetic acid)

Diethyl ether (ether)

Glycerol (which see)

Mono, di and tri acetin (combinations of acetic acid and glycerol)

Iso propyl alcohol (another type of alcohol, one of the congeners in alcoholic beverages)

Propylene glycol (converted to glycogen in the body).

SORBITOL

A sweetener used in diabetic foods and a humectant (see miscellaneous **Additives**). It is found naturally in small amounts in some fruits—for instance cherries—but commercial supplies are made from **glucose**.

Sorbitol, 60% as sweet as **sucrose**, is absorbed at a very slow rate into the blood stream during **digestion**. Though the precise way in which it is used by the body is uncertain sorbitol does not appear to raise the blood glucose level (see Diabetes) appreciably, and is a useful substitute for sucrose in diabetic diets. It is added to foods like diabetic jams, tinned fruits, chocolates, marmalade and sweets. However it cannot be used freely in slimming diets: unlike saccharin (which supplies no energy) sorbitol supplies 4 kilo**calories** per gram—the same as sucrose.

SOUPS

Perhaps surprisingly, soups are not especially nourishing: 75 to 95% of their weight is **water**. Their nutrient content depends on the ingredients used, but in general soups contain less than 5% **protein** and only small amounts of **minerals** and **vitamins**. 'Cream' soups usually contain more **energy** than other soups: clear soups are very low in energy. A clear soup, made from **extracts** or stock (left overnight in a refrigerator and the fat removed) with a few diced vegetables is suitable for slimmers. In sickness and convalescence, dried skimmed milk enriches vegetable soups. Broths are a good appetite stimulant, but of little value (see Gelatine) by themselves.

Canned soups, though standards are not controlled by law, are of similar value to home-made. Manufacturers follow a Code of Practice (not law, but it can be used as evidence in legal proceedings): meat soups must contain a minimum of 6% meat; scotch broth a minimum of 3%. Brown, Windsor and Eton soups (made from a brown roux and meat stock) and minestrone do not have to contain

APPROXIMATE NUTRIENTS IN 200 GRAMS (1 BOWL) OF SOME SOUPS

	Canned			Packet	
Nutrient	Tomato	Vegetable	Chicken	Spring Vegetable	Oxtail
Energy (kilojoules)	565	360	450	60	120
(kilocalories)	135	85	120	15	30
Protein (grams)	2	4	3½	1	2
Carbohydrate (grams)	19	16	10	3½	7
Fat (grams)	6	1	7	0	0
Sodium (milligrams)	965	1000	—	—	—
Potassium (milligrams)	415	400	—	—	—
Calcium (milligrams)	40	40	—	—	—
Iron (milligrams)	0·6	0·9	—	—	—
Other minerals	Small amounts		—	—	—
Vitamin A (micrograms)	100	—	—	—	—
Vitamin C (milligrams)	12	—	—	—	—
B vitamins:					
thiamin (milligrams)	0·06	—	—	—	—
riboflavin (milligrams)	0·04	—	—	—	—
nicotinic acid (milligram equivalents)	0·4	—	—	—	—
Other vitamins	—	—	—	—	—

— – no figures avilable

meat. Unless otherwise stated, consommé should be made from meat stock. In vegetable soups, a single vegetable (like tomato) should contain more of the named vegetable than others; mixed vegetable soup should contain at least four different vegetables; and green pea should be made from fresh or frozen peas. Cream soups must contain butter fat.

Dehydrated soups are not subject to regulations or a Code of Practice, though, like canned soups, their ingredients must be stated on the **label**. They tend to be lower in protein and energy than canned soups: some, especially the 'instant' kind, are a mixture of **additives—flavours, colours, emulsifiers**, and **preservatives** and **antioxidants** 'carried over' in ingredients—with little or no food value.

SOYA BEANS
See Pulses, Novel protein.

SPICES

Barks, flowers, seeds and roots of plants used to enhance and blend with the **flavour** of food and probably in the past to mask tainted meat and fish. They are eaten in tiny quantities and consequently are of no direct nutritional value (except red pepper and curry powder—see below). However they stimulate appetite and are part of good cuisine. Some, like cloves, cinnamon and mustard, have a preservative action.

Most spices contain no vitamins C or A. Red paprika and cayenne peppers however contain 30 to 60 vitamin A equivalents per gram, as **carotene**. Some of the colour in other spices is due to orange yellow pigments chemically related to carotene but with no vitamin activity—for instance curcumin in turmeric. Fresh horseradish contains 10 milligrams of vitamin C per 10 grams. Spices contain small amounts of B vitamins, calcium (0–10 milligrams per 1 gram) and other **minerals**. For instance 0·2–0·05 milligrams of iron per gram. Curry powder is a useful source of **iron**, containing 7·5 milligrams in an average portion of 10 grams.

Spice products. Curry powder should contain at least 85% spices and not more than 15% salt. Different blends are made from fenugreek with 3 to 4 of the following spices: turmeric, cinnamon, cassia, coriander, fennel, ginger, red pepper, pepper, mustard, allspice, cardamom, mace, cumin, dill and a piece of bay leaf.

Black pepper is the unripe fruit, white the ripe fruit with the outer coat removed. It is more pungent freshly ground, and black is more pungent than white. Mustards are a mixture of brown and yellow (black and white) varieties. If it has added starch it must be called mustard condiment or compound mustard. Prepared mustards (french, etc) are a mixture of the ground seeds with vinegar (or wine), salt, sugar and other spices.

Mixed spice contains 5 ground spices selected from cloves, ginger, coriander, fennel, cinnamon, cassia, red pepper and turmeric. Pickling spices are selected from whole coriander, allspice, chillies, ginger, white mustard, cloves, mace, cinnamon, black mustard and white pepper. Sausage seasonings contain sage, pepper, cloves, ginger, nutmeg, allspice, red pepper. Herbs, ginger, pepper, mace, allspice and cloves are used in delicatessen meats.

Spice extracts are the essential oils (see Flavour) dissolved in alcohol and preserved in sulphur dioxide or benzoic acid. Ginger and horseradish are also allowed to contain preservatives.

SPIRITS AND LIQUEURS

Spirits are concentrated sources of **alcohol**: they supply **energy** but are deficient in all other **nutrients**. A $\frac{1}{2}$ bottle of 70° proof spirit contains 840 kilo**calories**—nearly one-third of the **recommended intake** for energy for most men.

Spirits can be distilled from any fermented carbohydrate—for instance whisky and gin are made from cereals, vodka from rye or potatoes, brandy from grapes, rum from molasses, calvados from apples. Fusel oils (congeners) which distill across with alcohol contribute to the flavour. Fusel oils are responsible for some of the unpleasant side effects of alcoholic drinks, but are virtually removed during redistillation of vodka and gin. Juniper berries and coriander are added when gin is redistilled.

All spirits sold in Britain must be at least 65° proof (contain at least 65% of proof spirit) unless otherwise stated on the label. One hundred millilitres (about 3 measures) of 65° spirit contains 25 grams of alcohol, but most spirits are 70° proof containing 32 grams of alcohol per 100 millilitres. See Alcohol for conversion of proof into grams of alcohol.

Liqueurs contain more alcohol (30 to 40 grams per 100 millilitres) than spirits, are sweetened with sugar, and flavoured with herbs and spices. One measure (about 30 grams) contains 10 grams of **carbohydrate** and 100 kilocalories.

STARCH

Starch, a **carbohydrate**, is an important **energy** yielding constituent of food. Although starch is unnecessary for health, unrefined starchy foods are useful sources of essential **nutrients**.

Starch is found in seeds and some roots. Cereals contain about 70%, bread 50%, chestnuts 30%, potatoes 20%, pulses 10 to 15%, other nuts and root vegetables up to 5%. Leaves and fruits contain traces. Arrowroot, sago, tapioca, and cornflour are virtually pure starch. Pure starch, extracted from any starchy food, is extensively used by the food industry. On **labels** it is called 'edible starch'.

A diet containing 2 slices of bread for breakfast, a sandwich lunch and a supper with potatoes and a pudding contains about 150 grams, but individual intakes are very variable. In general, as income rises starch is replaced by increased consumption of sugar (**sucrose**) and **fat**. This trend may be harmful: unrefined starches also contain **protein**, **vitamins**, **minerals** and **fibre**, and a high fat, high sucrose diet is associated with an increased risk of **heart disease**. A **balanced diet**, and therefore good health, is more likely in a regime that includes starch in an unrefined form, as potatoes

and bread (preferably wholemeal) rather than one in which
potatoes and bread are avoided but foods high in fat and
sugar (confectionery, puddings, cakes) are eaten instead.
When low carbohydrate diets are prescribed (for disorders
of blood **lipids** and **diabetes**) all sugar should be avoided and
the daily allowance of carbohydrate taken mainly as unrefined starch and fruit.

Starch—a **polysaccharide** composed of many molecules
of **glucose** linked together in chains—is stored in characteristic granules in plants. Uncooked starch does not dissolve in water easily and is poorly digested. Grinding into
flour breaks down cell walls, but only partially damages
starch granules. When added to water starch takes up a
small quantity of water but does not dissolve until the mixture is heated, when the granules eventually burst (gelatinise)
thickening the food. Sauces are smoother and thicker if they
are stirred, which helps to rupture the granules. Dry heat
has a different effect—the starch is broken down to **dextrins**
(for example in toast, breakfast cereals). Prolonged heat
chars starch, rendering it useless for food.

Starch is dismantled during **digestion** by **enzymes** and the
liberated glucose absorbed into the blood stream. Starch
therefore yields the same energy (about 4 kilo**calories** per
gram) as glucose and other carbohydrates. Excessive quantities of energy from starch or other food will of course be
stored as fat and eventually result in **obesity**.

Pure (edible) starch is an important thickener and stabiliser (see Emulsifiers) in the food industry. For 'instant' foods
it can be cooked and dried (pregelatinised) so that it will
dissolve easily and thicken smoothly with little cooking. It
can also be modified in other ways (for example with acids,
enzymes or chemically altered) to conform with particular
needs of manufacturers. The Food Additives and Contaminants Committee have recommended that enzyme and
chemically modified starch should be brought under stricter
control—see Emulsifiers.

STARVATION—see also Marasmus

Severe loss of weight caused by insufficient food for bodily needs. It is usually associated with famine, but can be self-imposed, for example for political reasons or in anorexia nervosa, a serious mental condition affecting adolescent girls in particular. Loss of weight—caused by lack of appetite, fever, **malabsorption**, and some feeding regimes (for example low fat or protein diets) inadvertently low in **energy**—is also a feature of many other diseases.

Normal adults can survive with no food for about six weeks, thin people for less. As fat stores (adipose tissue) are depleted, energy expenditure is usually decreased (starving people are listless and apathetic) but sufferers from anorexia nervosa are often very active, giving a false impression of good health. The body proteins are also wasted away, first to supply essential **amino acids**, then, as fat stores are depleted, for energy. As muscles shrink, the skin becomes lax, though the accumulation of water may give a superficial well-fed appearance.

Few people can survive a loss of more than one-third of their normal body weight: death is usually precipitated by a heart failure or diarrhoea. In the early stages, recovery is complete once food is made available, but the outlook is less good if the digestive tract has degenerated. Specialist treatment, including very dilute feeds of skimmed milk and glucose or intravenous fluids are necessary until the digestive tract has regenerated. Anorexia nervosa cannot be treated by diet alone: there is an underlying resistance to gaining weight which can only be overcome with psychiatric care. Sufferers will go to great lengths to deceive—for instance they will force themselves to vomit or conceal food they are thought to have eaten and undertake vigorous exercises when left alone.

When weight has been lost after illness, small frequent meals are advisable until appetite has recovered sufficiently for normal meals (with extra helpings) to be taken. Fruit juices with glucose, or proprietary glucose and milk based drinks are useful ways of increasing energy intake. Treatment of the underlying disease sometimes requires a **therapeutic diet**.

SUCROSE—common name sugar

Commercially produced from beet, cane and to a lesser extent, maple. The chemical composition of sucrose is identical from all sources: each molecule of sucrose contains one of **glucose** and one of **fructose**. Most plants contain traces, but nearly all sucrose in the diet is derived from commercially produced white sugar; a highly purified substance containing 99·5% sucrose and no other nutrient. Sucrose supplies the same **energy** (approximately 4 kilo**calories** per gram) as other pure **carbohydrates**.

Pure crystals of sucrose (white sugar) are produced from cane or beet by a complex process of extraction. Different size crystals—caster, granulated or preserving—can be made by modification of the final processing. Icing sugar tastes sweeter because the ground powder dissolves quickly on the tongue. Demerara retains some of the colour and flavour of raw cane sugar, but it is little better than white as a source of nutrients. Other brown sugars are made by mixing refined sugar with syrup.

Molasses, which remains when sugar has been crystallised out of cane or beet sugar solutions, is used to make treacle and golden syrup. These contain minerals (see table on p. 320) and traces of B vitamins. Some of the iron and calcium are derived from machinery and lime used in purification. The table overleaf shows nutrient contents in 10 grams of sugars and syrups; one or two teaspoons of syrup or two teaspoons of sugar weigh 10 grams. Glucose (corn) syrup is made from starch, usually **maize**.

Sugar is a **preservative** and is necessary for 'lightness' in cakes, but valued primarily for its taste. After about 1850, it became a common article of food in Britain, and, apart

Syrup/sugar	Kilojoules (kilocalories)	Calcium milligrams	Iron milligrams	Potassium milligrams	Water grams
White	165(40)	0·1	0·004	0·2	0
Demerara	165(40)	5	0·08	10	0
Brown	145(35)	5	0·25	25	0·2
Molasses	105(25)	15 (light) 60 (black)	0·4 (light) 3 (black)	150	2·4
Treacle	105(25)	50	0·9	145	2·8
Golden syrup	120(30)	5	0·15	25	2

from restrictions during war and post-war shortages, consumption increased markedly up to 1956. In 1855, average total daily consumption per person was about 1¼ ounces, but 5 ounces in 1956. A similar pattern is emerging as underdeveloped countries become urbanised. A level of just over 2 lb total per person per week seems to be the saturation point, probably because then income has increased sufficiently for other, more satisfying foods, like meat and fat, to be bought. In Britain, daily consumption of sugar has been falling since 1956 to about 4 oz total sugar per person, which provides 16–18% of the total energy in the diet.

About half of the average total sugar is used in cooking and in sweetening tea and coffee. People who do not take sugar in drinks will of course eat less. Others, who take 2 teaspoons of sugar (10 grams) or more in drinks will exceed the average. The rest of the average intake is eaten in jam, sweets, cakes, biscuits, soft drinks and ice cream. Boiled sweets contain 90% sugar, toffees, jams and marmalades 70%, and chocolates 50%. Cakes and soft drinks vary.

During digestion, sugar is split to glucose and fructose which are absorbed into the blood stream and transported to the liver. In very large quantities fructose tends to be used in a different way from glucose, raising the level of some of the blood lipids. Very large quantities increase the blood cholesterol level in animals, but in men, women past the menopause, and some women taking the contraceptive pill, the blood triglycerides (or simple fats) are raised. The effect is not seen with normal quantities or if the diet is also high in polyunsaturated **fatty acids** (from sunflower seed oil), and low in saturated fats. It is likely to be one of the many risk factors in **heart disease**: a high intake of sugar alone probably does not affect mortality (some countries where consumption of sucrose is high, like Jamaica, Cuba, have a low rate of death from heart disease) but in conjunction with other risk factors—for example a diet high in saturated fats—the risk of heart disease is increased.

Heart disease apart, sugar is a food of poor nutritional value for people living sedentary lives. It contains energy but no other nutrient and can safely be discarded from the diet without risk of depleting the body of essential protein, vitamins and minerals. Sucrose is also known to be very harmful for teeth, particularly when it is eaten in a sticky form (like toffee), and may raise the level of triglycerides on the skin. The advice to those who suffer from a greasy skin and spots to avoid sugar is probably well founded.

SUGAR

Sugar is synonymous with **sucrose** but there are many other sugars in food. After sucrose, the most common are **glucose, fructose, maltose** and **lactose**. Mannose is found in manna—the gummy sap of the Tamarisk tree eaten in a sweet with nuts. They are all chemically related (as **carbohydrates**), converted to glucose in the body, and taste sweet. They have different sweetening powers: compared with sucrose, fructose is the sweetest (1·7 times) and lactose the least sweet (0·2 times).

Other substances that taste sweet but are not sugars are called artificial **sweeteners**.

SULPHUR

Part of the essential **amino acid methionine**, and of **cysteine** and **cystine**. Nearly all dietary sulphur is derived from **pro-**

teins containing these amino acids, though two B vitamins **(thiamin** and biotin) also contain small amounts. Apart from the needs for these essential nutrients, humans are not thought to require extra sulphur.

Adults contain about 120 grams of sulphur, mostly in proteins. Daily diets contain 600–1600 milligrams, depending on the quantity and quality of protein. An equivalent amount is filtered out daily from the blood by the kidneys, mostly as salts of sulphuric acid (sulphates).

Sulphates and other forms of sulphur—like sulphur dioxide—in food are probably treated in the same way by the body. Although its use is restricted, because it destroys thiamin, sulphur dioxide has been added to food for many years as a permitted **preservative**. However, its safety is under investigation: at extremely high dosages it has been found to cause genetic damage (see Poisons in food) to some bacteria, and it may react with fats when foods are stored for several months, causing toxic compounds. Most of the daily intake would be from wine and beer.

SWEETENERS

Substances which—like **sugars**—stimulate the sweet sensitive taste buds on the tongue. Artificial sweeteners however usually contain no **energy** and are much sweeter than sugars. Other substances in food also taste sweet, for instance glycerol and glycerine (0·7 times as sweet as sucrose), but they are not generally used solely as sweeteners.

Saccharin, 550 times as sweet as sucrose, has been in use for about 80 years. Each tablet contains an average of 13 milligrams of saccharin—an equivalent sweetening power to a heaped teaspoon of sugar. It is permitted in **soft drinks**— most may contain a maximum of 45 milligrams per pint—if it is declared on the label.

Saccharin is particularly valuable in diabetic and slimming diets, which must contain no sugar, and it also reduces the expense of soft drinks. However, its use in food is under review: two American studies in 1972 reported increased incidence of bladder cancer in rats fed large doses of saccharin. These findings have not been confirmed by several other studies, and may have been due to impurities.

It is possible that an acceptable daily intake (see Additives) of 5 milligrams saccharin per kilogram body weight may be proposed. For a 10-year-old child, the maximum daily intake would be about 150 milligrams per day, contained in three pints of pop. For an adult weighing 65 kilograms, the acceptable daily intake would be 325 milligrams or 25 saccharin tablets. It is probably wise to limit intake to this level—especially in hot conditions (when more perspiration is lost and the bladder contents are concentrated).

Apart from **sorbitol**, saccharin is the only artificial sweetener permitted in food. Cyclamate was banned rather hastily in 1970: American studies implicated a mixture of cyclamate and saccharin as a cause of bladder cancer in rats. Other alternatives are being investigated, in view of the bitter taste of saccharin (which intensifies after heating) and current doubts about its safety.

Aspartame, composed of two **amino acids** (phenylalanine and aspartic acid), is nearly 200 times sweeter than sucrose. It has recently been permitted in some foods in the USA and will probably be permitted in Britain. Monellin and thaumatin, two **proteins** extracted from berries, are 3000 times as sweet as sucrose. Like saccharin, these sweeteners are of limited use because they break down when heated. Other interesting substances are miracullin, which causes sour foods to taste sweet for several hours after being spread on the tongue, and maltol which does not taste sweet but enhances sweet fruity flavours. It is used in chocolate and essences.

TAPIOCA

Extracted from cassava (manioc) roots, partially cooked and sold as pearls (which resemble sago) or flakes. Untreated cassava from some sources is poisonous, containing cyanide.

The nutritional deficiencies of tapioca—it is virtually pure **starch**, yielding **energy** but few other nutrients—are overcome when it is cooked with milk. However, it is easily grown and has become the staple food in many communities, often where milk is in short supply. Children weaned on to cassava (but not given milk or other sources of protein) are at risk from the protein deficiency disease **kwashiorkor**.

AVERAGE NUTRIENTS IN 15 GRAMS
($\frac{1}{2}$ OZ) TAPIOCA

Energy (kilojoules)	225
(kilocalories)	55
Carbohydrates (grams)	14
Fat	Trace
Protein (grams)	0·06
Iron (milligrams)	0·05
B vitamins, nicotinic acid (milligrams)	0·03
Other vitamins and minerals	0 or traces only

TARTARIC ACID

Found in grapes, it is partly responsible for the acid taste of young wine. Little is absorbed from food. It is used to give a sharp taste to some soft drinks and jams. Cream of tartar and tartaric acid are the acid ingredients in some **raising agents**.

TEA

Originated in China and became a popular drink in Britain in the middle of the eighteenth century. Indian and Ceylonese supplies virtually replaced Chinese towards the end of the nineteenth century. The shoots are withered, rolled, fermented and dried. Green teas are not fermented.

Teas contain flavouring oils, **caffeine**, and tannin, an

astringent. Indian and Ceylon teas contain more tannin than Chinese and green teas contain more than black. Caffeine contents are about the same. Blends contain about 20 different types, selected for flavour and astringency. Caffeine is extracted out of tea more readily than tannin, so that first cups contain more of the stimulant and less of the astringent than second cups.

Tea is a good source of two **trace elements, manganese** and **fluorine**—and contains small amounts of two B vitamins, **riboflavin** and **nicotinic acid**. Other vitamins are largely destroyed during manufacture.

The table below shows the contents of a 150 millilitre (5 oz) cup of tea, made with 5 grams of tea and no milk.

Caffeine	Fluorine	Manganese	Riboflavin	Nicotinic acid
50–80 milligrams	0·2–0·5 milligrams	1 milligram	0·05 milligrams	0·3 milligrams

TEETH—loss of

Tooth decay (dental caries) is most active up to the age of 20. In adulthood, gum disease is more likely to cause loss of teeth.

Dental caries

The accumulation of bacteria in deposits (plaque) on the teeth is the most widely accepted cause of decay. The bacteria multiply on **carbohydrate** in food, producing **lactic acid** which dissolves out the protective enamel on the tooth. Unless the resultant cavity is filled, bacteria infect the inside of the tooth, causing toothache and eventual loss of the tooth.

Rarely, vitamin deficiencies before teeth erupt result in defective enamel, which is more easily eroded. Once the teeth have erupted they cannot be improved by extra vitamins or a high **calcium** diet. There is an unknown hereditary factor in tooth decay: some people have more resistant

AVERAGE NUTRIENTS IN 100 GRAMS (3½ OZ) PORTIONS OF SOME VEGETABLES

Nutrient	Beans, runner (boiled)	Beetroot (boiled)	Broccoli tops (boiled)	Brussels sprouts (boiled)	Cabbage, red (raw)	Cabbage (boiled)	Carrots (boiled)	Carrots (canned)	Cauliflower (boiled)	Celery (raw
Energy (kilojoules)	30	185	60	65	85	35	20	80	45	40
(kilocalories)	5	45	15	15	20	10	80	20	10	10
Protein (grams)	¾	1¾	3	2½	1¾	¾	½	¾	1½	1
Carbohydrate (grams)	1	10	½	1¾	3½	1¼	4¼	4½	1¼	1¼
Fat (grams)	0	0	0	0	0	0	0	0	0	0
Sodium (milligrams)	5	65	10	10	30	15	50	280a	10	135
Potassium (milligrams)	85	350	105	245	300	145	85	85	150	280
Calcium (milligrams)	25	30	160	25	55	60	35	25	25	50
Magnesium (milligrams)	15	15	15	10	15	5	5	5	5	10
Iron (milligrams)	0·6	0·7	1·5	0·6	0·6	0·5	0·4	1·3	0·5	0·6
Phosphorus (milligrams)	10	35	55	45	30	15	15	15	35	30
Other minerals	Small amounts									
Vitamin A (micrograms)	50	0	415	65	0	50	1500	1165	5	0
Vitamin C (milligrams) (1)	5	5	40	35	60	20	5	3	20	7
Vitamin D	0	0	0	0	0	0		0	0	0
Vitamin E (α tocopherol) (milligrams)	0·1b	—	—	1	—	0·5c	0·5b	—	0·15d	0·5
B vitamins:										
thiamin (milligrams)	0·03	0·02	0·06	0·06	0·06	0·03	0·05	0·04	0·06	0·3
riboflavin (milligrams)	0·07	0·04	0·2	—		0·05	0·03	0·04	0·02	0·06	0·3
nicotinic acid (milligram equivalents)	0·7	0·4	1·2	0·9	0·5	0·3	0·5	0·4	0·7	0·5
pyridoxine (milligrams)	—	0·05b	—	0·28b	0·12	—	0·1b	0·02	0·2b	0·1
pantothenic acid (milligrams)	0·05b	0·12b	1b	0·4b	0·18	—	0·25	0·1	0·6b	0·4
biotin (micrograms)	0·7b	Trace	—	0·4b	0·1	—	0·6b	1·5	1·5b	0·1
folic acid (micrograms)	—	—	100b	—	—	—	15b	—	35b	—
vitamin B₁₂	0	0	0	0	0	0	0	0	0	0

tooth enamel than others. Other **trace elements** (vanadium, molybdenum and selenium), besides **fluorine**, in water may affect the resistance of the enamel.

The most effective preventative measure against dental caries is a high standard of mouth hygiene. Teeth should be cleaned at least after breakfast and before bed. If there are no facilities for brushing teeth after other meals, the mouth should be rinsed with water. Faultily aligned teeth are more difficult to keep clean. Mouth breathing promotes plaque formation.

	Chicory (raw)	Cucumber (raw)	Egg plant or Aubergine (raw)	Leeks (boiled)	Lettuce	Marrow (boiled)	Mushrooms (raw)	Mustard and cress	Onion (raw)	Peppers, green (raw)	Parsnips (boiled)
Energy (kilojoules)	40	40	65	105	45	30	30	40	105	145	235
(kilocalories)	10	10	15	25	10	5	5	10	25	35	55
Protein (grams)	$\frac{3}{4}$	$\frac{1}{2}$	$\frac{3}{4}$	$1\frac{3}{4}$	1	$\frac{1}{2}$	$1\frac{3}{4}$	$1\frac{1}{2}$	1	2	$1\frac{1}{4}$
Carbohydrate (grams)	$1\frac{1}{2}$	$1\frac{3}{4}$	3	$4\frac{1}{2}$	$1\frac{3}{4}$	$1\frac{1}{2}$	0	1	$5\frac{1}{4}$	—	$13\frac{1}{2}$
Fat (grams)	0	0	0	0	0	0	0	0	0	0	0
Sodium (milligrams)	5	15	5	0·5	5	1	10	20	10	—	5
Potassium (milligrams)	180	140	240	280	210	85	465	335	135	185	295
Calcium (milligrams)	20	25	10	60	25	15	5	65	30	20	35
Magnesium (milligrams)	15	10	10	15	10	5	15	25	10	—	20
Iron (milligrams)	0·7	0·3	0·4	2·0	0·7	0·2	1	4·5	0·3	1	0·5
Phosphorus (milligrams	20	25	10	30	30	10	135	65	30	—	30
Other minerals	Small amounts										
Vitamin A (micrograms)	0	0	0	5	165	—	0	—	0	435	0
Vitamin C (milligrams) (1)	—	8	5	15	15	2	3	80	10	150	10
Vitamin D	0	0	0	0	0	0	0	0	0	0	0
Vitamin E (α tocopherol) (milligrams)	—	—	—	1[b]	0·05[f]	—	—	—	0·2	—	1[b]
B vitamins:											
thiamin (milligrams)	—	0·04	0·05	0·07	0·07	0	0·1	—	0·03	0·06	0·07
riboflavin (milligrams)	—	0·04	0·03	—	0·08	—	0·4	—	0·05	0·08	—
nicotinic acid (milligram equivalents)	—	0·1	0·8	—	0·4	—	4·5	—	0·4	0·7	—
pyridoxine (milligrams)	—	0·04	—	0·25[b]	0·07	0·05[b]	0·1	—	0·1	—	0·1[b]
pantothenic acid (milligrams)	—	0·3	—	0·12[b]	0·1	0·1[b]	2	—	0·1	—	0·5[b]
biotin (micrograms)	—	—	—	1·4[b]	0·7	0·4[b]	—	—	0·9	—	0·1[b]
folic acid (micrograms)	—	10	—	—	25	—	20	—	—	—	—
vitamin B_{12}	0	0	0	0	0	0	0	0	0	0	0

Toothpastes are helpful, but an effective and often renewed toothbrush is more important. Horizontal brushing is less effective in removing food particles and may also erode the enamel. Some toothpastes are too abrasive, removing the enamel with the stain. There is no necessity for fluorine-containing toothpastes in an area where the water has a good content.

Sugar (sucrose) is the main agent for promoting bacterial growth, especially when taken in a sticky form and between tooth brushing. The acidity at which enamel begins to be dissolved is reached within 5 minutes of eating sugar. The alkalinity of saliva protects to some extent. Apples, crisp

AVERAGE NUTRIENTS IN 100 GRAMS (*Continued*)

Nutrient	Radishes	Spinach (boiled)	Spring greens (boiled)	Swedes (boiled)	Tomatoes (raw)	Turnips (boiled)	Turnip tops (boiled)	Watercress	Avocado pears	Olives
Energy (kilojoules)	65	110	40	75	65	45	45	65	370	440
(kilocalories)	15	25	10	20	15	10	10	15	90	105
Protein (grams)	1	5	1¼	1	1	¾	2¾	3	1	1
Carbohydrate (grams)	2¾	1½	1	3¾	2¼	2¼	0	¾	2½	0
Fat (grams)	0	0	0	0	0	0	0	0	8	11
Sodium (milligrams)	60	125	10	15	5	30	5	60	15	2250a
Potassium (milligrams)	240	490	120	100	290	160	80	314	395	90
Calcium (milligrams)	45	595	85	40	15	55	100	220	15	60
Magnesium (milligrams)	10	60	10	5	10	5	10	15	30	20
Iron (milligrams)	1·9	4	1·3	0·3	0·4	0·4	3·1	1·6	0·5	1
Phosphorus (milligrams)	25	95	30	20	20	20	45	50	30	15
Other minerals	Small amounts									
Vitamin A (micrograms)	0	1000	—	0	115	0	1000	500	15	25
Vitamin C (milligrams) (1)	25	25	—	17	20c	17	40	60	20	0
Vitamin D	0	0	0	0	0	0	0	0	0	0
Vitamin E (α tocopherol) (milligrams)	—	2·5b	—	—	0·3	—	—	—	—	—
B vitamins:										
thiamin (milligrams)	0·04	0·07	—	0·04	0·04	0·03	0·06	0·1	0·1	0
riboflavin (milligrams)	0·02	0·15	—	0·03	0·04	0·04	0·2	0·16	0·1	0
nicotinic acid (milligram equivalents)	—	1·7	—	—	0·7	0·5	1·1	1·9	1·2	—
pyridoxine (milligrams)	0·1	0·1b	—	0·2b	0·1	0·11b	—	—	—	0·02
pantothenic acid (milligrams)	0·18	0·3b	—	0·11b	0·05	0·02b	—	0·1	—	0·02
biotin (micrograms)	—	0·1b	—	0·02	1·2	0·1b	—	0·4	—	—
folic acid (micrograms)	—	—	—	—	20	—	—	—	—	—
vitamin B₁₂	0	0	0	0	0	0	0	0	0	0

a – added during processing
b – figure for raw vegetable
c – raw cabbage heart—outer leaves contain more
d – none in flower
— – no figures available
(1) – boiled vegetables cooked in minimum amount of water for shortest possible time
e – canned tomatoes or juice, 16 milligrams vitamin C
f – raw lettuce heart, outer leaves contain more

foods and fibrous foods help maintain the health of the gums, but do not always prevent caries.

Very acid foods are capable of dissolving away the enamel: the front teeth are particularly affected. Acid cola and carbonated drinks are best taken with a straw, and other acid foods—citrus fruits, vinegar pickles, and acid drops—should not be retained in the mouth.

Gum Disease

Most adults, even with well cared for mouths, have some gingivitis—the most frequent cause of tooth loss in adult life. It is usually caused by accumulated plaque on the teeth. Bacteria invade the margin between the gum, causing inflammation and bleeding. If untreated, ulcers develop and the teeth loosen and fall out or have to be extracted. The teeth loosen and fall out in scurvy (deficiency of vitamin C), but this is rare.

Frequent effective tooth brushing—which removes plaque and stimulates the blood supply in the gums—is the most important preventive factor.

TEXTURED VEGETABLE PROTEIN

See Novel protein.

THERAPEUTIC DIETS

Normal diets altered—usually in their content of one or more **nutrients**, but also in their texture—for medical purposes. Because indisposition is often attributed to food, they were one of the first treatments used in medicine. Unfortunately many diets used in the past were based on unfounded beliefs, and some present-day claims for dietetic 'cures' are based on equally uncertain evidence. Nevertheless, therapeutic diets are an important part of modern treatments of many medical conditions.

Therapeutic diets are most commonly needed for weight reduction. *All* **slimming diets**, despite hundreds of 'different' regimes, are designed to contain less **energy** than normal food. No food has special slimming properties.

Fat, protein, carbohydrate, fibre, calcium, salt (sodium), potassium and some vitamin modified diets may be necessary for the treatment of many conditions, see individual entries. See also Coeliac disease, Diabetes, Inborn Errors of Metabolism, Gout, Heart disease, Dyspepsia, Ulcers, Allergy.

Substantial alteration of the diet carries with it the risk of unsuspected deficiency diseases—for instance, scurvy has occasionally resulted from some types of ulcer regimes. Self-treatment with very restricted diets is inadvisable—medical and qualified dietetic assistance should be sought.

THIAMIN—other name vitamin B$_1$
              —obsolete name, aneurine

Thiamin, part of the **vitamin B complex**, is essential for growth and life. The nervous system is first affected by a diet lacking in the vitamin, but deficiency diseases—**beriberi** and other nervous disorders—are now rare except amongst alcoholics.

Wholegrain cereals, pork, bacon, ham, heart, liver, kidney, cod roe, most nuts and pulses (peas and beans) are good sources of thiamin (see also Yeast). Most other fresh foods contain thiamin, but it is readily destroyed during cooking and processing (see below). Although it is removed in wheat germ (a rich source) and bran during milling, white flour is legally required to have added thiamin. Sugar, fats, refined starches and alcoholic spirits contain none.

Needs for thiamin depend on the amount of **carbohydrate** (sugar and starch) in the diet, but daily **recommended intakes** (twice the usual minimum needs) are related to **energy** requirements and are greater for men (1·1–1·4 milligrams) than women (0·9–1·0 milligrams). The table on p. 333 shows portions of foods supplying 0·9 milligrams, the intake for most women. Requirements are decreased when little carbohydrate is eaten, and may be increased by large intakes of alcohol. See Appendix for other recommended intakes.

Allowing for cooking and processing losses, the average British diet contains about 20% more thiamin than the

average reommended intake. Nearly half the average intake
is supplied by bread and other cereal products; vegetables
and meat contribute a further third.

Thiamin is vulnerable to heat, particularly when alkalies,
such as bicarbonate of soda, are used and—like other B
vitamins—leaches out of foods cooked in water. It is also
removed from refined cereals and is destroyed by sulphur
dioxide, a **preservative** permitted in some foods.

Alkalies and sulphur dioxide are the most destructive. In
soda bread, cakes and greens cooked with baking powder
or bicarbonate of soda, losses are increased proportionally
to the amount used. Bright green vegetables or yellow
scones (alkalies cause yellowing of flour) will contain no
thiamin. Self-raising flour and some bread mixes also contain bicarbonate of soda. To conserve thiamin, bread should
always be leavened with **yeast**. Sulphur dioxide is permitted
in some meat products that would otherwise be dangerous
sources of food poisoning (see Poisons in food)—like sausages, beefburgers and other meat with cereal products sold
uncooked—and (to prevent browning) in dehydrated vegetables and prepeeled chips and potatoes. These foods cannot
be relied on as sources of thiamin.

All cooking destroys some thiamin. Bread loses up to
30% when toasted for a minute, and at least 30% is lost
when meat is cooked. In general, grilled, fried or roast meat
loses less (30–60%) than stewed or braised meat (50–75%):
little is left in the gravy. Fish loses about 50%, peas, beans
and potatoes up to 30%, and roasted nuts up to 75%. Large
volumes of water increases losses in boiled vegetables by
up to 60%. Keeping food hot causes further destruction.

No thiamin is lost when meat and fish are frozen, but
losses can occur in storage: 20 to 40% can be lost when meat
is stored in a deep freeze for 6 months. Losses will be
greater if temperatures are allowed to rise above −18°C.
Canning causes losses of between 50 and 75%, and there are
further losses on storage if cans are not kept cool—up to
40% at warm (70°F) temperatures over a year. Canned

sardines and corned beef contain no thiamin. Less heat is
required to sterilise acid fruits: canned orange juice retains
most of the vitamin. Cured meats (which are preserved with
nitrite) retain most of the vitamin, but smoking causes a
20% loss. Bloaters and kippers contain no thiamin.

Thiamin is also destroyed during toasting and puffing
of breakfast cereals, which usually contain negligible
amounts unless it is added by manufacturers (when it will
be declared on the label).

The human body contains about 25 milligrams of thiamin,
fairly evenly distributed throughout all cells. As part of a
co**Enzyme** thiamin is necessary for many metabolic (see
Metabolism) processes, most importantly the release of
energy from **glucose**. Nervous tissue is especially dependent
on glucose as a source of energy and is markedly affected
by a diet containing minimal amounts after about 3 weeks.
Initial symptoms are loss of weight, due to nausea and vomiting, sleeplessness, depression, irritability, and failure to
concentrate. Later changes include degeneration of the
nerves and atrophy of the muscles they supply—causing
loss of sensation and weakness, particularly in the legs, and
loss of memory—and sometimes enlargement of the heart
and accumulations of body fluids.

Beriberi is never seen in healthy adults in Britain, although
mild deficiency in the elderly may exacerbate confusion and
loss of memory. However, in alcoholics, deficiency is common and manifests itself either as beriberi, or heart failure
or severe lesions in the brain. If treated early enough the
damage can be reversed, but sometimes the brain and
nerves are irreversibly damaged, resulting in permanent
confusion, impairment of memory and difficulty in walking.
Underlying disease, such as **malabsorption**, cancer and vomiting in pregnancy may also cause deficiency.

Thiamin cannot be stored and excesses as pills or tonics
are filtered out of the blood stream by the kidneys. Although
often prescribed as a tonic in the past, thiamin alone is now

thought to be of limited value: a diet needing supplements of thiamin is also likely to be lacking in other B vitamins and occasionally large doses of thiamin have precipitated other B vitamin deficiencies. A **balanced diet** is a better safeguard of health than vitamin pills and tonics.

THIAMIN CONTENTS OF FOOD

Food		Average milligrams thiamin per 100 grams food	Approximate grams containing 0·9 milligrams	Measure
Rich	Cod roe	1·5	60	
	Wheat germ	1·25	70	
	Brazil nuts	1·0	90	2 to 4 ounces
	Fresh peanuts	0·9	100	
	Roast pork (a)	0·8	110	
Good	Oatmeal	0·5	180	
	Fried bacon, braised heart	0·4	220	6 to 10 ounces
	Kidney, liver	0·3	300	
	Boiled peas (a), roasted peanuts	0·25	370	14 ounces
	Wholemeal bread (b)	0·2	450	11–12 large slices
	White bread (b)	0·18	500	
Moderate	Eggs	0·1		
	Oranges, raisins, avocado pears	0·01		Between one and two kilograms
	Boiled potatoes (a)	0·08		
	Roast beef	0·05		
	Milk	0·04		Between two and three kilograms
	Cheddar cheese	0·04		
	Boiled cabbage	0·03		

(a) Sulphur dioxide is added to hamburgers, sold uncooked, prepeeled potatoes and chips (used in most restaurants and canteens) sausages and (if declared on the label) dehydrated peas and vegetables. At permitted levels, about half the thiamin in meat may be destroyed (subsequent frying leads to total destruction) and probably all in vegetables.

(b) At least 20% lost with baking powder or soda. Nearly 90% destruction with excessive quantities (sufficient to yellow the flour).

THREONINE

An essential **amino acid** needed in the daily diet for replacement (repair) of body **proteins** lost through daily wear and tear, and for new proteins made during growth. All food proteins supply threonine: the estimated daily adult needs are supplied by, for example, the 11 grams of protein in 330 grams (just over $\frac{1}{2}$ pint) of milk. Excesses not needed for new body proteins are converted to glucose and used for energy purposes. The **nitrogen** part is converted to urea, later filtered out of the blood stream by the kidneys.

TIN

Probably an essential **trace element** for animals. If it is also essential for humans, the normal daily intake of 1 to 4 milligrams would be more than adequate. Canned foods and foods wrapped in tin foil contain more than unprocessed foods.

Excess tin in foods causes a metallic taste and can be toxic. Legally, foods should not contain more than 250 parts per million (2·5 milligrams per 100 grams of food). Some cans are lined with lacquer to prevent erosion by acid contents. Other foods do not cause erosion in unopened cans, but once in contact with the oxygen in air, tin is rapidly dissolved. Food should not be stored in open cans.

TRACE ELEMENTS

Minerals needed in tiny amounts in the diet for health. So called because older methods of analysis could detect only unmeasurable traces in food and living tissues. With modern methods of chemical analysis the needs for trace elements are becoming known more precisely, especially for animals, but accurate estimations of quantities in foods are still lacking.

Ten trace elements are presently known to be essential nutrients for humans: iron, iodine, copper, zinc, manganese, cobalt, molybdenum, selenium, chromium, and vanadium. Silicon, tin, nickel and fluorine are essential nutrients for animals, but it is not known if they are necessary for humans. Many other elements occur in traces in the body, food and water, but are not thought to be essential for life or health.

Excess quantities of trace elements are poisonous and there is often only a small difference between the quantity needed for health or safety and the quantity which is toxic. Relative excess of some trace elements can cause deficiency of others. The risk of obtaining too little or too much in food is minimised by eating a wide variety of foods in moderation (see Balanced diet).

Food processing, changes in agricultural practices (like excessive use of fertilisers and new breeds of plants) and changes in dietary habits (like eating more fat in place of cereals) may alter the balance of trace elements in the diet. These alterations may affect the development of some degenerative diseases, for instance **heart disease**. Cadmium (see Poisons in food), calcium, chromium, copper, selenium, vanadium and zinc are some of the trace elements currently under investigation for their possible relationship to heart disease.

TRIGLYCERIDES

Simple **fats** (sometimes also called neutral fats). Each molecule of triglyceride contains one of **glycerol** and three **fatty acids**. In food, triglycerides always have different fatty acids. The hardness (melting temperature) and other characteristics of a fat are determined by its content of fatty acids.

Triglycerides can be modified in many ways to make **emulsifiers** and stabilisers. The simplest modified fats, superglycerinated fats, contain less than 3 fatty acids for each molecule of glycerol (these are also formed in **digestion**). Glycerol monostearate, for example, contains one fatty acid (stearic acid) for each molecule of glycerol. Other emulsifiers have organic acids (like tartaric, lactic, and acetic acids) or **sorbitol** substituted for fatty acids, or glycerol.

TRYPTOPHAN

An essential **amino acid**, needed in the daily diet for the rebuilding (repair) of **proteins** lost through daily wear and tear, and for new proteins made during growth. If the diet contains sufficient **pyridoxine** tryptophan is converted to **nicotinic acid** and serotonin—a potent regulator in the brain and most other parts of the body.

Tryptophan is limiting in **maize** protein, but most other food proteins contain adequate amounts. The estimated adult needs are supplied by, for example, the 25 grams of protein in 310 grams (8 large slices) of bread, or the 15 grams of protein in 140 grams (2 large) eggs.

TYROSINE

An **amino acid** made in the body from phenylalanine. It is required for the formation of new **proteins** needed for growth and repair and is the precursor of some hormones (like the thyroid hormones) and the brown pigment melanin formed in hair, eyes and tanned skin. Tyrosine is found in all food proteins and reduces the requirement for phenylalanine. Excess quantities are converted to **glucose** or **fat** and used for **energy**: the **nitrogen** part is converted to urea, later filtered out of the blood stream by the kidneys.

High levels of tyrosine—due to a temporary insufficiency of an **enzyme** necessary for its normal **metabolism**—sometimes accumulate in the blood stream of babies. The disorder is made worse by lack of vitamin C (necessary for the action of the enzyme) and artificial milk (cows milk contains more phenylalanine and tyrosine than breast). Vitamin C supplements, as orange juice or welfare vitamin drops, minimise the possibility of resultant brain damage. Permanent deficiency of the enzyme—hypertyrosinaemia, a rare **Inborn Error of Metabolism**—can cause liver and kidney failure unless treated with a synthetic diet low in phenylalanine and tyrosine.

Foods containing tyramine, a derivative of tyrosine, must be avoided when certain tranquillisers are taken—see Cheese.

ULCERS (peptic)—see also Digestion

Open sores in the walls of the digestive system in contact with digestive (peptic) juices secreted by the stomach. They usually occur in the duodenum (which connects the stomach to the small intestine), sometimes in the stomach itself (gastric ulcers) and rarely at the bottom of the oesophagus.

Peptic juices contain two potent substances—hydrochloric acid and pepsin, a protein splitting **enzyme**—and it is surprising that ulcers do not occur more commonly. Normally however the walls of the stomach and duodenum are resistant to erosion: a sticky mucous substance is also secreted and probably acts as a barrier. The underlying cause is unknown, but in duodenal ulcers, worry, stress and overwork are important predisposing factors. There is also a familial tendency and men and people in blood group O are more commonly affected. Gastric ulcers are thought to occur less frequently now: this decrease may reflect improvements in the general standard of nutrition over the past 30 years.

Pain—when acid comes into contact with the raw surface—from gastric ulcers is usually felt shortly after eating, and from duodenal ulcers when the stomach is empty. The pain is relieved by milk or other antacids (alkalies that neutralise the acid in the stomach) and a bland diet (see Dyspepsia). Most ulcers heal spontaneously, but rest in bed and some drugs accelerate healing. Though occasionally necessary as part of hospital treatment if an ulcer haemorrhages (penetrates a blood vessel) complicated ulcer dietary regimes are now thought to have little effect on healing. Those prone to ulcers should however avoid irritants like smoking, **caffeine** and alcohol in excess (particuarly when no food is eaten) and highly spiced foods. Rest, both before and after small meals (avoiding fried foods), spaced regularly through the day, is also advisable.

Ulcers that perforate or do not heal are surgically removed, which sometimes results in **malabsorption**.

VALINE

An essential **amino acid** needed in the daily diet for the rebuilding (repair) of body **proteins** lost through daily wear and tear, and for new proteins formed during growth. It is found in all food proteins: the estimated daily adult needs are supplied by, for example, the 14 grams of protein in 50 grams (nearly 2 oz) of roast beef. Excesses eaten, not needed for growth and repair are converted to glucose and used for energy purposes. The nitrogen part is converted to urea, later filtered out of the blood stream by the kidneys.

Valine is restricted in the treatment of Maple Syrup Urine Disease, see Isoleucine.

VANADIUM

A **trace element** only recently found to be an essential nutrient for some animals and thought to be essential for man, although food contents and normal intakes are uncertain. Supplements of vanadium affect blood **lipids** and the rate of dental decay in animals, but the relevance of these findings to human **heart disease** and dental decay is unknown.

Excess vanadium in fumes or dust from certain industrial processes can be harmful to personnel, but there is little risk of toxic levels in food.

VEGETABLES—see Table pp. 326–8

Sources of **vitamins, minerals** and **fibre**, vegetables can be leaves, stems, roots, flowers or fruits of plants. Dark green leaves are superior sources of most nutrients. Potatoes, maize (sweetcorn), pulses (peas, beans, lentils) are discussed under separate entries.

All fresh vegetables are sources of **vitamin C**, but it rapidly declines during storage, cooking and processing. Leaves, tomatoes and peppers are better sources than other vegetables—10 to 30 grams of raw green peppers, sprouts, or broccoli tops fulfils the daily adult recommended intake for vitamin C. In the average diet, the daily intake of 4 ounces of vegetables (excluding potatoes and pulses) supplies about 25% of the total intake of vitamin C.

Green and yellow/orange vegetables are sources of **vitamin A** (as carotene). Carrots and dark green vegetables (kale, cress, broccoli) are the richest sources: 100 grams (3½ oz) supplies at least half the daily adult recommended intake. Other root vegetables (including beetroot) and white vegetables, like cauliflower and celery, contain none. In the average diet, vegetables supply about 20% of the total intake of vitamin A.

Fresh vegetables are low in **energy**: 70 to 95% of their weight is water, and most are fat free and low in **carbohydrate**. In **slimming diets**—which usually allow plenty of vegetables—salads are no better than cooked vegetables, though perhaps more palatable. Avocado pears and olives may have to be avoided: they contain fat with a low proportion of polyunsaturated **fatty acids**. Carrots, beetroot and parsnips contain 5–10% of carbohydrate—slightly more than other vegetables.

Vegetables contain small amounts of **protein**, and are sources of **vitamin E**, but, apart from **folic acid** are at best only moderate sources of B vitamins. Mushrooms however are good sources of **pantothenic acid**. All contain **potassium**, and dark green vegetables are better sources of iron, copper and calcium than others. Spinach, beet greens and chard contain **oxalic acid** which renders most of their calcium unavailable for absorption into the blood stream.

All vegetables contain fibre (marrow and cucumber the least), mostly in the form of **pectin** and **cellulose**. Although cooking may not reduce fibre contents, cellulose is softened and pectin partially degraded, and nutrients inside cells become more accessible for digestion. Vegetables should still have 'bite' when cooked: overcooked vegetables are unpalatable and will have lost most of their vitamin C (see below).

Vitamin C is the nutrient most vulnerable to storage, cooking and processing. It leaches out of foods cooked in water and is rapidly destroyed by heat, alkalies, oxygen, iron, copper and the enzyme (ascorbic acid oxidase) released from cells damaged by peeling and chopping. The table on p. 342 shows how to keep losses to a minimum. Losses of folic acid are proportional to those of vitamin C. Minerals and vitamins E and A however are fairly stable.

Vitamin C begins to decline as soon as vegetables are harvested. Any damage, like wilting or bruising destroys it, and the losses are accelerated in warm temperatures. Greens can lose up to 50%, beans 20% in a day when kept at room temperature. Wilted lettuce leaves are likely to contain no vitamin C. Losses are minimal when whole vegetables are kept covered in a cool place, preferably in the salad drawer of a refrigerator, but there can be severe losses if vegetables are allowed to be frost bitten. In salads, vitamin C decreases immediately vegetables are sliced, chopped and dressed.

Losses of vitamin C are greatest when vegetables are cooked in a large volume of water: about 40% is lost from greens cooked in a little water at the bottom of the pan, and 80% if they are completely covered. Most of the losses remain in the vegetable water and can be recovered if used immediately (but not if stored—see below). Pressure cooking and steaming cause the least loss—10 to 30%—and minerals are well retained. About half the potassium is lost when vegetables are boiled.

Assuming losses of 40% of vitamin C, a 100 gram portion of home cooked cabbage may contain about 40 milligrams of vitamin C—enough to fulfil the adult daily recommended intake. However, large-scale cooking usually results in losses of 80%: cabbage served freshly cooked in a canteen may only contain 12 milligrams per 100 grams. Bright green or soft vegetables (indicating that bicarbonate of soda has been used) will contain none. Keeping hot causes further loss (about 60% every hour), and total losses may be 90%, when cabbage will only contain 5 milligrams per 100 grams. Pre-cooking, storing in a refrigerator, and reheating portions as they are wanted (thus avoiding the necessity to keep vegetables hot) also causes loss of vitamin C (20 to 30% every 24 hours). Vegetables eaten in canteens and restaurants are unreliable sources of vitamin C.

Before processing all vegetables are blanched (dipped in hot water to reduce bacterial contamination and to inactivate enzymes that would otherwise cause loss of vitamin C and flavour) and this causes losses of 10 to 30% of vitamin C. However, vegetables are blanched when their vitamin content is at a maximum and blanching losses may be less than those encountered in transport and storage in shops.

Freezing does not affect vitamin C, but it gradually declines during storage. There is hardly any loss at very low temperatures ($-30°C$) but at normal deep freeze temperatures ($-18°C$), up to 30% can be lost over a year. Finely divided foods—like spinach—will lose more (up to 50%). There are severe losses at higher temperatures (at $-12°C$, 50% can be lost in 4 months). Vegetables should never be thawed before cooking and must always be cooked according to the manufacturers instructions, otherwise there will be further losses: vegetables that have been allowed to defrost (shown by ice crystals in the pack) should be rejected. Vitamin A is stable to freezing, but vitamin E may decline during storage.

Canned vegetables have to be sterilised by heat, causing further loss of vitamin C. Overall, most canned vegetables lose between 40 and 60%, but acid tomatoes retain most of the vitamin. Canned foods should be stored in cool (10 to 15°C) conditions, when they retain most of their vitamin C. Losses are accelerated in warm temperatures—up to 15%

can be lost over a year when cans are stored at 27°C. Canned vegetables must be reheated according to manufacturers' instructions and must not be stored in open containers otherwise all vitamin C may be lost. There may be severe losses of vitamin E on canning but vitamin A is stable.

Dehydrated vegetables that have retained their colour will probably have retained half their vitamin C, but losses are very variable. Sulphur dioxide (a **preservative** declared on **labels**) preserves vitamin C, but destroys thiamin. Dehydrated vegetables kept cool (10°C) lose little vitamin C, but there can be severe losses at higher temperatures (up to 50% in four months at 30°C). Carrots lose 20 to 40% of vitamin A when dehydrated.

Other methods of preserving vegetables are pickling and

CONSERVATION OF VITAMIN C IN VEGETABLES

Agents and processes that destroy vitamin C	To conserve vitamin C
Loss during storage	Use fresh foods when possible. Keep foods in a cool dark place. Do not allow to wilt
Waste during preparation	Prepare immediately prior to cooking
Heat from (a) cooking	Boil cooking water before putting vegetables in the pan. Do not cook whole vegetables – slice immediately before cooking. Small pieces take less time to cook. Cook for the shortest possible time; vegetables should have 'bite' after cooking
(b) keeping hot	Serve vegetables immediately they are cooked
Leaching into water during	
(a) cooking	Use the minimum amount of water for cooking
(b) washing	Use as little water as possible and do not allow to soak. Wash intact leaves quickly
Enzyme released from damaged cells after frost, bruising, pounding or chopping	Enzyme inactivated at 65°C. Slice and peel vegetables with a sharp knife, then drop into boiling (100°C) water immediately
Oxygen in air	Do not leave vegetables to stand or soak. Boil water to expel air before putting vegetables in to cook. Use a lid on the pan to keep air out. Do not store cooked vegetables or liquor – even in a refrigerator
Alkali in bicarbonate of soda	Do not use. Lemon juice and vinegar are acids and will help to preserve vitamin C
Iron from rusty implements	Do not use
Copper from copper pans	Do not use

fermenting. Sauerkraut, preserved in salt and **lactic acid** formed by bacterial action, is a very variable source of vitamin C, but most samples can be expected to contain half the vitamin C of properly cooked cabbage. Other pickles are preserved in salt and **vinegar**, sometimes with added lactic or acetic acids, and the permitted preservatives sulphur dioxide or benzoic acid. Thick pickles are a mixture of vegetables, spices, sugar, salt and vinegar, thickened with **edible gum** or starch. They can be coloured, for example brown with caramel, yellow with tartrazine. Pickles are of little nutritional value, except as aids to appetite.

Tomato sauce however contains small amounts of vitamins A and C. Tomato sauce must legally contain no other vegetables except tomato (6% minimum), onion and spices. Colour and preservative are permitted, but must be declared on the label. Tomato paste (minimum of 28% tomato) is a more concentrated source of tomato than purée (minimum 11% of tomato).

VEGETARIANISM

Avoidance of meat and fish, but not dairy products or, usually, eggs.

Nuts, pulses and dairy products supply the same range of essential **nutrients** as meat and fish and most vegetarians are able to eat a **balanced diet**. They may also be less susceptible to **osteoporosis**. However, meat and fish are the best sources of **iron** and **iodine** and vegetarian diets may supply too little for women, children and adolescents. These difficulties are best overcome by eating soya products (the next best source of well absorbed iron) and iodised salt or seaweed products.

Vegan diets, which exclude all animal products, lack **vitamin B$_{12}$** and incur the risk of pernicious **anaemia**.

VINEGAR

Contains **acetic acid**, a strong acid which is an effective **preservative**. Acetic acid is formed from alcohol by bacteria: vinegar is made from malted barley (malt vinegar); grape juice or wine (wine vinegar); cider; or spirits. Colourless vinegars are usually distilled from malt vinegar. Vinegar should not contain less than 4% acetic acid and may contain caramel, to colour brown, and sulphur dioxide (a preservative). Non-fermented vinegars, made from synthetic acetic acid, must be called 'non-brewed condiment'.

Vinegar has no slimming properties, but it contains virtually no energy and is therefore permitted freely in slimming diets. Cider, wine and malt vinegars contain traces of B vitamins and minerals.

VITAMINS—see also Nutrient

Organic (that is made by bacteria, plants or animals) substances needed in minute quantities for life, growth and health. They cannot be made in the human body in sufficient amounts to satisfy needs, and must be eaten in food.

Humans are definitely known to require 13 different vitamins, which have different functions and occur in different foods:

VITAMINS KNOWN TO BE ESSENTIAL FOR HUMANS

Vitamin	Chemical name	Discussed under
A	Retinol and carotene	Vitamin A
B	Eight different	See vitamin B complex
C	Ascorbic acid	Vitamin C
D	Cholecalciferol and ergocalciferol	Vitamin D
E	Tocopherol and tocotrienol	Vitamin E
K	Phylloquinone and menaquinone	Vitamin K

Choline, para amino benzoic acid, inositol, carnitine and lipoic acid are essential nutrients for micro-organisms and some animals, but not for man. The **bioflavonoids** (including rutin) are not proven essential nutrients for micro-organisms, animals or man.

Vitamins are sometimes divided into fat soluble (A, D, E, and K) and water soluble (B and C) groups. Although vitamins A and D were first isolated from dairy (fatty) foods, the division is largely of historical interest: E and K occur in fruits and vegetables. However, fat soluble vitamins are more stable to cooking and processing than vitamins B and C. Water soluble vitamins leach out of foods into cooking water and most are readily destroyed by heat. They are consequently more vulnerable to cooking and processing.

Although it had been known for some time that certain foods were protective against diseases like scurvy and beriberi, the existence of the vitamins was not established until 1912 by F. G. Hopkins. They were then gradually isolated from foods, lettered, and—after identifying their chemical structure—given a chemical name. Vitamins F, G, H, L, M and P are now obsolete names: they were later found not to be vitamins or to be misidentifications of substances which had previously been discovered. See also **Vitamin B complex**.

VITAMIN A—chemical name, retinol

Vitamin A maintains the health of the skin; is necessary for colour and night vision; and is probably involved in the senses of taste and balance.

Liver is the richest source of vitamin A. Other good sources are margarine, eels, eggs and dairy products: milk and cream contain slightly more in summer than winter. Carrots, and dark green vegetables are also good sources: they contain **carotene**, a precursor of vitamin A. Fatty fish, other offal, molluscs (like mussels), green and yellow vegetables and fruit also contribute small amounts. In the average British diet (excluding meals taken outside the home) about 25% of the total intake is supplied by butter and margarine,

25% by liver, 20% by vegetables, and 20% by other dairy foods and eggs. Cereals, nuts, meat, white fish, white vegetables and fruits, oils (except red palm oil), and white fats contain virtually no vitamin A. Fish liver oils, which used to be used as dietary supplements are very rich sources.

The daily adult **recommended intake** for vitamin A—more than twice the average minimum needs to allow for individual variation—is 750 microgram equivalents. The table below shows portions of foods that will meet the adult daily intake. Women need more throughout breast feeding to ensure adequate supplies for the growing child, and children need more in proportion to their weight than adults to allow for growth. See Appendix. The average British diet supplies more than the average recommended intake—1270 microgram equivalents in 1973.

AVERAGE PORTION OF FOODS SUPPLYING THE RECOMMENDED ADULT
DAILY INTAKE OF VITAMIN A

	Food	Portion supplying an average of 750 microgram equivalents of vitamin A
Rich sources	Liver, ox	5 grams ($\frac{1}{6}$ oz)
	calves	5 grams ($\frac{1}{6}$ oz)
	lambs	5 grams ($\frac{1}{6}$ oz)
	chicken	10 grams ($\frac{1}{3}$ oz)
	pigs	10 grams ($\frac{1}{3}$ oz)
	Old carrots	40 grams ($1\frac{1}{2}$ oz)
	Eels[a]	60 grams (2 oz)
	New carrots, spinach, butter margarine	75 grams ($2\frac{1}{2}$ oz)
Good sources	Cheddar cheese, double cream, broccoli	180 grams (6 oz)
	Eggs	250 grams (4 standard)
Moderate sources	Tinned apricots	450 grams (1 lb)
	Milk	2000 grams ($3\frac{1}{2}$ pints)
	Herring	1650 grams ($3\frac{1}{2}$ lb)

a – based on an average content of 1200 micrograms per 100 grams, but may vary from 250 to 2500 micrograms

Vitamin A used to be measured in international units (i.u.), which for historical reasons were defined in units of carotene. To simplify matters, in 1969 the Department of Health and Social Security recommended that vitamin A should be expressed as vitamin A (retinol) microgram equivalents. British i.u.s can be converted to equivalents by multiplying by 0·3. Margarine, for instance, which contains a minimum of 760 i.u.s per ounce, contains 228 (760 × 0·3) microgram equivalents per ounce. Equivalents, used throughout this book, are gradually replacing i.u.s. Retinol is sometimes called vitamin A_1 to distinguish it from another form, vitamin A_2 (dehydroretinol), which has only half the potency of retinol and is rarely eaten (it is found in the liver of some fresh water fish).

Vitamin A is fairly stable to heat, and does not leach out of foods cooked in water. There are minimal losses in cooking and processing, but the vitamin is sensitive to light and oxygen, and is destroyed by rancid fats. **Antioxidants** added to most fats prevent rancidity and loss of vitamin A, but fish liver oils left in clear glass bottles are likely to be less potent than those kept in dark glass. See also Carotene.

Adults eating the recommended intake have usually accumulated about 150,000 micrograms of vitamin A, of which 90% is stored in the liver. In the eye, vitamin A is attached to a protein in the cells of the retina (light sensitive part) responsible for perceiving dim and red or green light. Light slightly alters the structure of the vitamin A molecule, stimulating nerve receptors in the cells. An impulse is transmitted to the brain, which registers light. The vitamin may act in a similar way in other sense organs responsible for smell, taste and balance. One of the first symptoms of vitamin A deficiency is an inability to see in dim light (night blindness) but there are many other causes—including anxiety—which are not affected by vitamin A.

The eye contains less than 10 micrograms of vitamin A, but paradoxically the precise function of the remaining 15,000 micrograms (excluding the liver) in the body remains uncertain. It is however known to be essential for the integrity of the skin and other 'linings' (epithelial tissues) which are the body's main defence against infection. Once the liver stores are exhausted by a deficient diet, after the onset of night blindness, cells of the skin, and cornea of the eye become hard and flake off, plugging the lubricating sweat and tear ducts. The skin becomes dry and may develop a typical 'toad skin' appearance, but permanent gooseflesh (folliculosis) often seen in well nourished people is not due to vitamin A deficiency. The cornea of the eye becomes dry and wrinkled, and eventual infection and resultant scarring cause permanent blindness.

In animals deficient in vitamin A the bronchial tubes become blocked with cells from the lining of the tubes, causing broncho-pneumonia. Animals also appear to need vitamin A for reproduction, the development of sound teeth and prevention of some types of kidney stones and possibly cancers, but the relevance of these findings to human health is uncertain.

Vitamin A deficiency, aggravated by lack of protein, is very common in poverty stricken communities. In India alone, 10,000 children are reported to become blind every year through lack of vitamin A, though in Britain, most people have sufficient vitamin A held in store for at least several months. Even the early signs of deficiency—night blindness—are very rare occurring only where there is underlying disease like **malabsorption** (the vitamin is transferred from food to the blood stream dissolved in fat). Nevertheless many elderly and middle-aged people are thought to have low liver stores. Those who do not like liver, and who need to avoid fats for medical reasons may not fulfil the recommended intake. A daily portion of dark green vegetables or carrots (failing this one vitamin A supplement, see below) should be taken. It is particularly important that the recommended intake should be met in pregnancy: a mother's inadequate diet may check the baby's growth and possibly cause brain damage. Supplements of vitamin A (for example welfare vitamin drops) are necessary for baby feeding.

Although by maintaining the health of the skin, vitamin A does prevent infection, supplements are of no value in protecting against or curing colds. Claims that vitamin A supplements are a 'natural' protection against sunburn are based on very dubious experiments: no controls (given placebo tablets) were used and the reported limited successes could have been due to other factors. Vitamin A is dangerous in excess. Except under medical advice, no more than one daily supplement should be taken. Over dosage (when 100 times or more the recommended intake is given) results in poor appetite, loss of hair, headache and peeling itching skin. The symptoms disappear when the vitamin is stopped, but in children excessive vitamin A may cause permanent deformities of bones. Intoxication never arises from diet alone (unless polar bear, whale, or shark liver is eaten—polar bear liver contains 60,000 micrograms per 10 grams) but teaspoons of halibut oil and vitamin pills have caused poisoning in children. The death of a man from liver failure in 1974 caused by a daily intake of 2,000,000 micrograms of vitamin A was given wide publicity.

SUPPLEMENTS OF VITAMIN A

Supplement	Dose	Vitamin A content in microgram equivalents
Col liver oil BP	1 teaspoon	900
Halibut liver oil BP	1 drop	900
Malt with cod liver oil	60 grams (1 dessertspoon)	1350
Welfare vitamin drops for babies	7 drops	300
Haliborange capsules	1 capsule	750
Multivite tablets (a)	1 tablet	700
Welfare mothers' vitamin tablets	1 tablet	1200

(a) others vary; vitamin potencies listed on box (100 i.u.s = 30 micrograms).

VITAMIN B COMPLEX

All occur in yeast, cereal germ and liver (except vitamin B_{12} which is only found in animal foods), but also in varying amounts in other foods including meat. Eight are known to be essential nutrients for human health (see table below) and all have different chemical structures and different (though related) functions in the body.

B complex vitamins known to be essential nutrients for humans

—see individual entries:

Thiamin (B$_1$)
Riboflavin (B$_2$)
Pyridoxine (B$_6$)
Vitamin B$_{12}$
Nicotinic acid
Folic acid
Biotin
Pantothenic acid·

Because the B complex vitamins occur in the same types of foods, a shortage of one B vitamin alone is unlikely. Most multivitamin pills contain only some—often the cheapest— of the complex. Since B vitamins also complement each other's action, an improvement of diet (see Balanced diet) is preferable to tonics and pills. Except in some medical disorders (see individual entries) excesses above the recommended intakes serve no useful purpose: they are eliminated almost immediately in the urine.

Originally the germ of cereals was thought to contain only one vitamin—the anti beriberi vitamin or vitamin B. Others were later discovered and numbered vitamins B$_2$ to vitamin B$_{12}$. However, some were later found not to be vitamins or had already been discovered. Consequently only four B complex vitamins are still numbered—B$_1$ thiamin, B$_2$ riboflavin, B$_6$ pyridoxine, and vitamin B$_{12}$.

VITAMIN B$_{12}$—group name for several forms of the vitamin, including cyanocobalamin and hydroxycobalamin

Part of the **vitamin B complex** necessary for protection against pernicious anaemia. It is plentifully supplied in animal foods, but almost entirely absent from plants: vegans (who eat no animal foods) are likely to become deficient, but there are many medical causes of pernicious anaemia.

Although liver is the richest source, all animal foods—heart, kidney, fatty fish, shell fish, meat, cheese, white fish, eggs and milk—supply vitamin B$_{12}$. Most British diets supply at least 4 micrograms each day, which is probably more than double the adult requirement. The table opposite shows portions of foods supplying 2 micrograms. Women need more than usual during pregnancy to ensure adequate supplies for the child.

Vitamin B$_{12}$ is fairly stable to heat, but, like other B vitamins, leaches out of foods cooked in water. Up to half can be lost when meat is cooked, but little is known of the losses incurred in food processing.

During **digestion** vitamin B$_{12}$ is transferred from the small intestine into the blood stream. Intrinsic factor—a substance made in the lining of the stomach—is essential for adequate absorption of the vitamin. Once in the blood stream, the vitamin is stored in the liver or taken up by cells and converted to coEnzymes involved, with **folic acid**, in the replication of cells during division. Very active cells, continuously dividing, need more than others. The vitamin is also needed for growth and to maintain the integrity of the insulating (myelin) sheath of nerves, and possibly the reproductive system.

Liver stores tide the body over when insufficient vitamin B$_{12}$ is absorbed from food. Most people have stores to last for at least 2 years, but they are eventually exhausted, when blood levels fall and pernicious **anaemia** develops. The bone marrow produces fewer blood cells and they are immature. The disease causes pallor and tiredness, the tongue becomes bright red and occasionally diarrhoea develops. More seriously nerves lose their myelin sheath and degenerate, causing a variety of symptoms, including mental disturbance and spasticity.

Pernicious anaemia mostly develops in middle age, probably as a result of an inherited defect in the immune system.

No intrinsic factor is formed and in its absence, vitamin B$_{12}$ cannot be absorbed into the blood stream. Capacity to absorb the vitamin may also be lost following surgery (removal of parts of the stomach or small bowel); infections (fish tapeworm or bacteria in the small intestine sequester the vitamin); and some drug treatments. The disease used to be fatal but is now treated with injections of vitamin B$_{12}$ (formerly raw liver) which cause a remarkable improvement in well being.

Vegans, who eat no animal foods—unlike vegetarians who usually eat milk, cheese and sometimes eggs—are also likely to become deficient unless vitamin B$_{12}$ tablets are taken. Folic acid—plentifully supplied in vegan diets—can

FOODS CONTAINING VITAMIN B$_{12}$(1)

	Food	Portion containing 2 micrograms vitamin B$_{12}$
Rich sources[a]	Liver, ox kidney, brislings, pilchards, sardines, herring	20 grams ($\frac{2}{3}$ oz) or less
Good sources	Other kidney, tongue, turkey, tuna, salmon	20 to 100 grams (3$\frac{1}{2}$ oz)
	Beef, lamb, veal, egg	100 to 200 grams (7 oz)
Moderate sources	Cheese, white fish, chicken, ham, pork, milk, tripe	200 grams to 650 grams (1$\frac{1}{2}$ lb)

(1) Based partly on values given by Adams et al (1973) B.J. Nut. 29 65
a – brown seaweed is also reported to contain vitamin B$_{12}$

mask the anaemia but not prevent gradual degeneration of the nervous system which is sometimes irreversibly damaged. Some vegans appear to escape the disease and it has been suggested that soil bacteria or fungi contaminating root or fermented vegetables may supply small quantities of vitamin B$_{12}$.

Diets temporarily lacking in animal protein foods are not harmful, provided the needs for other essential **nutrients** are met. Large doses of vitamin B$_{12}$ are not toxic: little of the vitamin is absorbed into the blood stream.

VITAMIN C—chemical name **ascorbic acid**

Small amounts of vitamin C in the diet are essential to allow growth in children and to protect against the deficiency disease **scurvy**. Requirements are increased during recovery from surgery or severe accidents, but it is controversial whether large intakes are otherwise conducive to improved health and resistance to infection.

Fruits and vegetables supply almost all the vitamin C in the diet. Cereals, eggs, fats, and most dairy foods, fish, nuts and meat contain none. Exceptions are liver, kidney, fresh fish roe, some oysters, unripe walnuts and some cured meats (like tinned ham) which have ascorbic acid added during curing. **Milk** and freshly killed meat contain a little, but it quickly disappears during storage. In the average British diet, nearly a quarter of the total intake is derived from **potatoes**, a quarter from other vegetables and one-third from **fruits**.

Of the fruits, citrus are the most reliable source of vitamin C, followed by berries. Other rich sources are blackcurrants and rosehip syrups. Apples, plums, peaches and pears contain little, dried fruits none. Of the **vegetables**, green peppers, sprouts, cresses and dark green leaves are the best source, but overcooked vegetables may contain none by the time they are eaten (see below). Root and blanched vegetables (like celery) contain little: dried (not dehydrated) vegetables contain none, but it is formed in intact **pulses**— peas, beans and lentils—allowed to sprout.

The British daily recommended intake for adults for vitamin C is 30 milligrams. More vitamin C is required during pregnancy and breast feeding to ensure adequate supplies for the child. See Appendix. The table on p. 356 shows average portions of foods containing sufficient to meet the adult allowance, but it is important to remember that vitamin C contents are greatly affected by the type and variety of plant grown, soil, climate and maturity when harvested. Green peppers for instance can vary five fold.

Storage time and method of cooking however, cause the most important variations in vitamin C contents of fruits and vegetables.

Fresh fruits and vegetables contain more vitamin C than those that have been stored because the soft tissues of harvested plants are alive and continue to use up the vitamin C accumulated during growth of the plant. Thus, the vitamin C content of potatoes falls from 30 milligrams per 100 grams when harvested to 10 milligrams per 100 grams after 6 months' storage. Losses are increased if vegetables are allowed to wilt, or if they are damaged by frost or bruising.

Vitamin C is the nutrient most vulnerable to cooking and processing. It leaches out of foods cooked in water and is rapidly destroyed by oxygen (in air); heat (especially if alkalies, iron or copper are also present); and the **enzyme** (ascorbic acid oxidase) released from vegetable cells during peeling and chopping. Some or all of these are in contact with vitamin C when food is cooked, and losses are unavoidable. For this reason, citrus fruits and tomatoes, which can be eaten without preparation are the most reliable sources.

There are further losses when vegetables are kept hot after cooking. Vegetables eaten in canteens and restaurants are unreliable sources and in these situations, tomato or cress salads are a better alternative. Lettuce (expecially if wilted) and cucumber, chicory and celery contain little vitamin C in an average portion.

The proportion of vitamin C lost during processing and preservation depends on the amount of heat to which the food is subjected. Acid fruits lose less than vegetables. A comparison between processed foods and foods freshly prepared in the home shows, perhaps surprisingly, that differences in vitamin C may be minor. In some cases preserved vegetables may contain more vitamin C than stored, unpreserved ones because they are processed when the vitamin C content is at a maximum.

The three most common methods of preservation are freezing, canning and dehydration, all of which begin by blanching (dipping in hot water) when 10 to 30% of the vitamin is lost. There is no further loss on freezing, but canned foods are sterilised by another period of heat when more vitamin C is destroyed. Dehydrated foods are also subjected to heat, but in modern processes, losses of vitamin C are minimal. Foods that have been dried by prolonged exposure to heat or sun will contain none. Processed foods should be stored at the lowest temperature possible and cooked according to the manufacturers' instructions, otherwise further losses of vitamin C may occur. Frozen vegetables for example should be cooked from the frozen state, and fruit juices should not be stored once the can or bottle is opened. See Pulses, Potatoes, Vegetables and Fruit for more information.

Nearly all animals studied synthesise sufficient vitamin C for daily needs from **glucose**. Humans and guinea pigs are two exceptions: they lack one of the **enzymes** in the vitamin C synthesising chain and must obtain the vitamin from food. Within about 4 months, a diet totally lacking in vitamin C results in the breakdown of connective tissue, a cementing substance that binds cells together (the vitamin takes part in the formation of the main proteins in connective tissue). The symptoms of scurvy—bruising (from leakage of blood into surrounding tissues), loss of teeth, weakening of bones —are due to loss of binding substance between cells. If given in time vitamin C restores full health; otherwise, sudden death is likely from internal haemorrhage. For most adults, a daily intake of 10 milligrams of vitamin C will prevent scurvy, but recommended intakes (30 milligrams for most adults) are greater to allow for individual variation. It is claimed that some women are able to synthesise small quantities of the vitamin and are better able to withstand scurvy.

Requirements for vitamin C are increased by wounds— incurred during surgery, burns or severe accidents which are repaired by scars, a form of connective tissue. Increased requirements are temporarily met by withdrawal from the blood, but these reserves are small. Hospital diets and intravenous fluids often contain insufficient vitamin C and supplements may be necessary for adequate healing. Babies may also develop scurvy unless given vitamin C (as welfare vitamin drops or orange juice, not boiled). See also Tyrosine.

Recommended intakes are more than adequate to prevent scurvy, but it is often proposed that—for full health—humans require additional vitamin C. These proposals stem partly from animal experiments and partly from uncertainties about other roles of vitamin C. It is implicated in many body processes (including brain activity, formation of red

AVERAGE PORTIONS OF FOOD SUPPLYING THE ADULT RECOMMENDED
INTAKE OF VITAMIN C

Food	Average portion supplying 30 milligrams vitamin C
Rich sources	Blackcurrants, strawberries, rosehip syrup, blackcurrant syrup, watercress, mustard and cress, green peppers, pea shoots, broccoli tops,[a] brussels sprouts,[a] cabbage,[a] cauliflower,[a] horseradish,[a] kale,[a] parsley,[a] spinach,[a] turnip tops[a]	15 to 50 grams ($\frac{1}{2}$ to $1\frac{3}{4}$ oz)
Good sources	Oranges, lemons, grapefruit, tangerines, redcurrants, gooseberries, blackcurrant, jam, bean sprouts,[a] new potatoes,[a] broad beans,[a] beet greens,[a] seakale,[a] turnip tops,[a] asparagus[a]	50 to 100 grams ($3\frac{1}{2}$ oz)
	Blackberries, limes, melon, pineapple (fresh), raspberries, canned grapefruit, avocado pears, runner beans,[a] leeks,[a] lettuce, spring onions, parsnips,[a] peas,[a] radishes, swedes,[a] tomatoes, turnips,[a] potatoes[a] up to 6 months old, liver, roe, some cured meats	100 to 200 grams (7 oz)
Poor sources	Apples, apricots, bananas, cherries, grapes, peaches, pears, plums, canned pineapple, rhubarb, most jam, marmalade, artichokes,[a] beetroot, carrots,[a] celery, cucumber, endive, marrow,[a] mushrooms,[a] onions,[a] potatoes older than 6 months, milk	200 grams to 1 kilogram ($2\frac{1}{4}$ lbs)

Processed peas, dried pulses (peas, beans, lentils), dried fruits (currants, dates, figs, prunes, raisins, sultanas, peaches, apricots), olives and jams made with fruits poor in vitamin C contain none.
a – There are considerable losses in cooking, see vegetables

blood cells and cholesterol synthesis) and may be essential for normal metabolism of all cells. Needs may be increased by stress. However, apart from its role in the formation of one hormone, its fundamental action remains elusive. Until this problem is resolved it will be difficult to establish whether or not raised intakes are beneficial to otherwise healthy people.

Vitamin C cannot be stored, but large doses are retained in the body until cells are saturated. A fully saturated 70-kilogram (11-stone) man contains about 5 grams of vitamin C. Once the cells are saturated intakes in excess of needs are eliminated in the urine. For full health, guinea pigs need enough for saturation, but most authorities remain unconvinced that recommended intakes should ensure saturation of human tissues. In Britain the recommended intake would then be unrealistically high: 60 to 100 milligrams daily are necessary for saturation but the average British diet only supplies 50 milligrams. Averages also mask individual intakes and many people in apparent good health habitually take even less than the present recommended intake.

Even larger doses—2 to 10 grams daily (equivalent to the quantities synthesised by animals not dependent on food for vitamin C)—are claimed to enhance resistance to infections from the cold virus, and from bacteria which later infect the damaged membranes of the sinuses, nose and lungs. A group of white blood cells, the body's main defence against infection, are less efficient when they are depleted of vitamin C, and their vitamin C content decreases with the onset of a cold. This fall can be prevented by a large dose (6 grams, or 1 teaspoon of powdered vitamin C) taken immediately symptoms are noticed, though not by a regular daily dose of 1 gram of vitamin C. Nevertheless, experiments to test whether or not people recover from colds more quickly if they are given the very large (6 gram) doses of vitamin C have had equivocal results: some have borne out the reputed benefits, others have not. Small doses (30 to 50 milligrams) contained in most cold remedies are unlikely to confer any benefits. Smokers have low white cell levels of vitamin C and are possibly more susceptible to infections: however smoking per se does not cause scurvy. Similar pharmacological doses (far greater than could be obtained from food) are also used in the treatment of schizophrenia: other diseases claimed to respond to vitamin C are diabetes and heart disease.

Up to 500 milligrams of vitamin C is probably a harmless daily intake, but larger doses should be taken with care. **Oxalic acid** formed partly from vitamin C is excreted in urine; excess amounts passing through the kidney may result in kidney stones.

VITAMIN D—chemical names cholecalciferol, ergocalciferol Vitamin D—obtained from food and formed in the skin exposed to sunlight—is necessary for the absorption of **calcium** from food, and the hardening of bones with calcium and **phosphorus.** Insufficient Vitamin D causes **rickets** in childhood and **osteomalacia** in adulthood.

Vitamin D is confined to very few foods. The only notable sources are fatty **fish, margarine** and **eggs.** Dairy foods (butter, milk, cheese) and liver contain very small quantities. Other foods (fruits, vegetables, cereals, other fish, fats, oils and meat) contain none. Fish liver oils, which used to be used as dietary supplements, are rich sources. Vitamin D is stable to heat and does not leach out of foods cooked in water. The only losses thought to occur in cooking and processing are with fat when milk is skimmed and fatty fish are grilled.

The **recommended intake** for vitamin D up to the age of 5 and during pregnancy and breast feeding is 10 micrograms per day. The table opposite shows portions of foods supplying this amount. The elderly and other people who live a confined life probably need at least 2·5 micrograms daily (the provisional adult recommended intake) in food. This is contained in, for example, 30 grams (1 oz) margarine.

Amounts of vitamin D formed in the skin cannot yet be accurately measured and other adult requirements are very uncertain. People who enjoy sunbathing or who spend much time outdoors may not need any dietary supplies.

There are several forms of vitamin D, but only two— vitamin D_2 (ergocalciferol) and D_3 (cholecalciferol)—are important to humans. Vitamin D_3 is the 'natural' vitamin found in food and formed in the skin. Vitamin D_2 is made when yeast is treated with ultra-violet light: it is used for

SOURCES OF VITAMIN D

	Food or supplement	Quantity supplying an average of 10 micrograms vitamin D
	Cod liver oil	1 teaspoon
	Welfare vitamin drops for children	7 drops
	Welfare vitamin tablets for mothers	1 tablet
	Other vitamin tablets[a]	
Rich sources	Fatty fish:	
	eels	10 to 40 grams ($\frac{1}{3}$ to $1\frac{1}{3}$ oz)
	brislings	40 to 20 grams ($\frac{1}{3}$ to $\frac{2}{3}$ oz)
	herrings, kippers	45 grams ($1\frac{1}{2}$ oz)
	mackerel	55 grams (2 oz)
	salmon fresh or canned	80 grams ($2\frac{3}{4}$ oz)
	sardines canned	135 grams ($4\frac{3}{4}$ oz)
	Cod liver oil and malt	45 grams (1 dessertspoon)
Moderate sources	Margarine	125 grams ($4\frac{1}{2}$ oz)
	Eggs	665 grams (11 standard)
	Fortified dried milk	115 grams (4 oz)
	Fortified evaporated milk	345 grams (12 oz)
Poor sources	Butter	800 grams (28 oz)
	Liver, cheese, cream, unfortified dried, evaporated and condensed milk, milk	More than 1 kilogram ($2\frac{1}{4}$ lbs)

a – potencies listed on box—100 i.u. = 2·5 micrograms

vitamin D supplements (for example in pills and margarine) and is thought to be as potent as D_3 for humans (though not for some animals). Vitamin D is measured in micrograms (millionths of a gram), used throughout this book, but potencies used to be assessed as international units (i.u.s). One i.u. of vitamin D is equal to 0·025 micrograms. Margarine for instance, which must legally contain 80 to 100

i.u.s per ounce, contains an average of 90 × 0·025, which
equals 2·25 micrograms per ounce.

Dietary vitamin D is transferred with fat into the blood
stream after digestion. In the skin, a derivative of cholesterol
is transformed to vitamin D by ultra-violet light from sunlight or ultra-violet lamps. The two sources are either stored
(probably mostly in adipose tissue—'fat') or converted to
active hormone-like forms. There are probably several
active forms, but one—made partly in the liver and later in
the kidney—stimulates the production of a protein necessary
for the absorption of calcium out of food. When little of the
skin is exposed to sunlight, coupled with insufficient vitamin
D in the diet, little calcium is available to the body. See
Rickets and Osteomalacia. Diseases which interfere with the
absorption of fat (malabsorption), kidney failure, congenital
diseases, and some drugs used for the treatment of epilepsy
can also cause rickets and osteomalacia.

Although most adults probably derive sufficient from sunlight, extra vitamin D—for example as welfare vitamin
tablets—is necessary in pregnancy and breast feeding, otherwise calcium is withdrawn from the mother's bones to meet
the needs of the growing child. With increasing age the kidneys may not convert vitamin D to its active form so efficiently: osteomalacia is therefore also likely in elderly people.
Substitution of margarine in place of butter and extra calcium (in at least ½ pint of milk a day) is advisable, together
with one multivitamin tablet daily for those confined in-
doors.

To prevent rickets, both breast and artificially fed babies
need vitamin D, as welfare vitamin drops or cod liver oil.
After weaning, vitamin drops, cod liver oil (or malt with
cod liver oil) should be continued up to the age of 5:
children and adolescents infrequently exposed to sunlight—
for example Asians and children confined to high rise flats—
are likely to benefit from supplements until they stop growing.

Excessive vitamin D is toxic. As little as 10 times the

recommeneded allowance has caused ill health in some
infants. Extra calcium is absorbed from food, resulting in a
dangerously high level of calcium in the blood stream. Death
may result from kidney failure following the deposition of
calcium in the blood vessels. For this reason, no more than
the recommended dose of vitamin drops or pills (or no more
than a daily teaspoon of cod liver oil) should be given to
children or taken in pregnancy. Adults are less susceptible
to overdosage than children: doses in excess of 25 times the
recommended intake have proved toxic.

VITAMIN E—group name for several substances (the
tocopherols and tocotrienols) in food and
synthetic forms

Known to protect animals against a variety of disorders—
including sterility and a type of muscular dystrophy—but
in adult humans there are only mild abnormalities specifically attributed to lack of vitamin E.

Nearly all foods contain vitamin E, and, because of its
widespread occurrence, deficiency is rare. Although soluble
in fat and found in fatty foods (vegetable oils are the richest
source) vitamin E is also supplied by foods containing little
fat, like cereals, fish, fruit and vegetables. The table overleaf
shows vitamin E contents of some foods.

There are probably minimal losses of vitamin E in cooking
—it is fairly stable to heat and does not leach out of foods
cooked in water—but there may be severe losses in processing and storage. For example, although wholemeal bread
is a good source of vitamin E, much is removed, with the
germ, during milling. Bleaching agents, used in white **flour**
destroy any remaining vitamin. Refined breakfast cereals
also contain little or no vitamin E. In fats, vitamin E is an
antioxidant, preventing rancidity and preserving vitamin A,
but eventually it is oxidised and made inactive itself. Rancid
fats and oils will contain none. Synthetic antioxidants added
to most fats protect against rancidity, but the vitamin can

SOME SOURCES OF VITAMIN E (1)

Food	Approximate[a] content in milligrams α tocopherol per 100 grams food
Rich sources Wheat germ, most vegetable oils	10 to 50 milligrams
Olive oil, peanuts	5 to 10 milligrams
Good sources Eggs, butter, cheese, cream, ox liver, wholemeal flour and bread, oats, brussels sprouts, leeks, parsnips	1 to 2 milligrams
Wholewheat breakfast cereals, apples, plums, carrots, celery, peas, cabbage	0·5 to 1 milligram
Poor sources White flour and bread, rice (white and brown), refined breakfast cereals, most meat and fish, milk, oranges, grapefruits, bananas, strawberries, gooseberries, lettuce, tomatoes, rhubarb, cauliflower, onion	Less than 0·5 milligrams

a – Foods vary greatly in their content of vitamin E: although some is removed in refining, refined oils may contain more than crude, because of natural variation. Vegetables can vary by at least 6 fold, depending on the part of the plant eaten and season (slow growing vegetables are reported to contain more than fast growing).

(1) Partly based on values given by Herting (1969), *J. Agric. Fd. Chem.*, 17, 785; Slover (1971), *Lipids* 6, 291; Slover, *Cereal Chem.*, 46, 635; Booth and Bradford (1963), *Brit. J. Nut.*, 17, 575; Herting (1963), *J. Nut.*, 81, 335; Bunnell et al (1965), *Am. J. Clin. Nut.*, 17, 1; and Bieri et al (1975), *J. Am. Diet. Ass.*, 66, 135.

be lost from retail butter, which is not permitted to contain antioxidant. Repeatedly used frying oils will also contain none. Little is otherwise known of the stability of vitamin E to processing: canned vegetables are reported to lose up to 90%, and there may be severe losses during prolonged deep freeze storage.

There are at least 8 substances in food which have vitamin E activity. Potencies (expressed as international units, i.u.s) are assessed from experiments with rats—the animal first found to be susceptible to lack of vitamin E. α tocopherol is the most potent: other forms, although they contribute to the total vitamin E contents of diets and are antioxidants in food, are much less potent—at least for rats. The vitamin can also be synthesised: dl α tocopherol acetate is the synthetic stable form commonly added to food and vitamin E tablets. One milligram of dl α tocopherol acetate is equal to 1 i.u. of vitamin E.

Adults probably need a minimum of 3 milligrams of α tocopherol daily, but British authorities feel unable to specify reliable **recommended intakes**. There are uncertainties about the potencies of the different forms of vitamin E for humans and lack of extensive information about vitamin E contents of normal diets. Attempts to establish requirements are made difficult by the rare occurrence of deficiency and further complicated by other constituents of food. Poly-unsaturated **fatty acids** increase requirements for vitamin E and **selenium** and sulphur amino acids (like **cystine**) may decrease them. Allowances of at least 12 i.u. per day for American diets—which contain more polyunsaturated fatty acids—are almost certainly too high for most Britons. The average British daily diet probably does not contain more than 5 milligrams of α tocopherol.

Vitamin E is held in the membranes of cells, and is likely to be important in maintaining their structure (precise organisation of molecules within cell membranes is essential for normal **metabolism**) but in adults there are surprisingly few obvious harmful effects of deficiency, apart from low blood levels and rather less robust red blood cells. Its basic function remains the subject of much speculation—at one time it was assumed to act principally as an antioxidant, preventing damaging alteration by oxygen of molecules of **lipid** in cells, but this explanation has been criticised. Several years of severe restrictions are necessary to deplete the body of the considerable store accumulated by adulthood. Diseases which interfere with the absorption of fat (during digestion Vitamin E is transferred to the blood stream dissolved in fat) sometimes result in similar mild symptoms of deficiency.

Babies—especially prematures born with inadequate stores—are more susceptible to deficiency of vitamin E. The resultant anaemia (due to increased destruction of red blood cells) and accumulation of body fluids are remedied by supplements of vitamin E. Supplements may also prevent the disease retrolental fibroplasia, which causes blindness and is more common in premature than full-term babies.

Based on the fact that similar diseases occur in animals made deficient in vitamin E, large doses—200 milligrams or more daily—of α tocopherol have been tried in the treatment of many conditions, including **atherosclerosis**, habital abortion, sterility and muscular dystrophy. Despite adequate blood levels before treatment (and therefore absence of deficiency) some doctors claim the vitamin to be of benefit to their patients. However, most doctors have found no benefits. Large doses are also given to athletes. The assumed antioxidant effect of vitamin E is believed to aid muscular efficiency—but this is unlikely. The performance of schoolboy swimmers given supplements of vitamin E for several weeks was no better than that of boys given placebo tablets of exactly the same appearance. Equally well-controlled experiments would probably confirm that extra vitamin E does not revitalise sexual powers, nor that it prolongs life or removes wrinkles. It may however accelerate wound healing (from burns or accidents) in people who have low blood levels.

Compared with vitamins A and D, vitamin E is said to be harmless in large doses for man, but animals are reported to become infertile. 100 milligrams daily (30 times minimum needs) is thought to be safe for those who believe in the reputed benefits of vitamin E.

VITAMIN K

Vitamin K is necessary for the formation of proteins responsible for clotting of blood. In its absence, prolonged bleeding can lead to death, but humans are not entirely dependent on dietary supplies. It is synthesised by bacteria inhabiting the digestive system, and at least half the daily requirement is thought to be obtained from this source.

Although a fat soluble vitamin, most vitamin K in the diet is supplied by vegetables: other foods (apart from liver) contain hardly any. Adult daily requirements—about 100 micrograms—are easily met by a helping of vegetables. Cabbage, sprouts, cauliflower, and spinach are probably the richest sources, containing up to 4 milligrams (4000 micrograms) in each 100 grams. Beans, peas, potatoes, carrots and liver contain up to 400 micrograms per 100 grams. Losses of vitamin K in cooking and processing are thought to be minor: it does not leach out of foods cooked in water and is relatively stable to heat.

In food, the vitamin has a slightly different structure (vitamin K_1, called phylloquinone or phytomenadione) from that made by bacteria in the digestive system (vitamin K_2, called menaquinone). Putrid food, infected with bacteria, also contains vitamin K_2. Many other forms have been manufactured, including K_3 (called menadione) which is more potent than K_1 and K_2. Natural forms are absorbed out of food dissolved in fat, which must be finely divided (emulsified) by bile salts before it can be transferred to the blood stream. Once in the blood stream, vitamin K is transported to the liver, where blood clotting proteins are formed. Antagonists of vitamin K—like Warfarin—are used medicinally to inhibit unwanted blood clotting. It is uncertain whether vitamin K has any other functions in the body.

Newborn babies are susceptible to vitamin K deficiency: in the first few days of life they have no bacteria inhabiting the digestive system, and milk is a poor source. Supplements are sometimes necessary to avoid death from haemorrhage. Adult deficiency is unlikely, except when the bile duct is blocked by gall stones—a cause of jaundice. The body is quickly depleted of vitamin K (unlike the other fat soluble vitamins there are no large stores) and injections are necessary before operations to remove stones, otherwise there is a risk of prolonged bleeding. Other causes of **malabsorption** can also result in deficiency.

Natural forms of vitamin K are not thought to be toxic when taken in excess, but some synthetic forms are soluble in water. They are useful for the treatment of vitamin K deficiency (being absorbed in the absence of bile) but in

amounts greater than 100 times the minimum requirement
can cause brain damage.

WATER

Though not strictly a **nutrient**, water is essential for life. In
temperate climates, most adults need to drink at least $1\frac{1}{2}$
pints (5 cups) of water or **beverages** daily. In hard water
areas, drinking water is a significant source of **calcium**—
London water for instance contains 200 parts per million
of calcium (or about 200 milligrams in an average intake of
1 litre). Hard waters are associated with a lower risk of
mortality from **heart disease**, but the reason for this is unknown. Calcium or other minerals in hard water may be
protective, or there may be harmful factors in soft water—
which is more acidic and tends to dissolve lead and other
unwanted trace elements (see Poisons in food) out of pipes
and containers.

Perhaps surprisingly, over half (60%) of the weight of an
adult man is water. Most is held inside cells: relatively little
(about $4\frac{1}{2}$ pints in a 70-kilogram (11-stone) man) is in the
blood stream. Women contain less water than men: they
have more fat which is drier (10% water) than muscle (80%
water). Babies contain more water than adults, and the
elderly slightly less.

Water is needed, from drink and food, to balance losses
of water in urine and from the skin, lungs and faeces. Normally losses are almost exactly balanced by gains; the table
below shows an approximate water balance for adults.

APPROXIMATE WATER BALANCE FOR ADULTS LIVING IN A
TEMPERATE CLIMATE

Gains (millilitres)	are balanced by	Losses (millilitres)
1000 in drink		1000 in urine
1000 in food		900 in skin
300 in metabolism		400 in faeces and lungs
2300	Total	2300

In temperate climates, drinking water usually balances the water needed by the kidneys for elimination of waste products in urine. A minimum of 900 millilitres (1½ pints) is usually required, but more in old age and infancy (when the kidneys are unable to concentrate waste products as efficiently) and if the diet is high in **protein** and **salt**. Requirements for drinking water are also increased in hot conditions and feverish illnesses to make good losses, necessary to cool the body, in perspiration.

In cool conditions, losses of water from faeces, lungs and skin (losses from the skin continue even when there is no obvious perspiration) are balanced by gains from food. Daily diets supply at least 750 millilitres of water. Milk, fruit and vegetables contain the most (75 to 95%). Meat, fish and eggs contain 50 to 75%; cheese and bread 40%; flour, cereals and nuts 5 to 15%. Sugar and fats contain little or no water. Metabolic water (released when fats, carbohydrates and proteins are used for energy) and water in soups, gravies and sauces also contribute to the total.

When losses are greater than gains, body fluids are concentrated, triggering a thirst response in the brain. Normally the desire to drink prevents dehydration when water is freely available, but the elderly may suppress thirst for fear of incontinence. Babies—who cannot ask for water—are also susceptible to dehydration.

When thirst is not satisfied, losses (a minimum of 1 litre a day in adults) continue. A loss of 2 litres is harmless but may contribute to confusion in the elderly. When more than 4 litres is lost however, water is withdrawn from cells, causing weakness and collapse. Few people would survive a loss of more than 8 litres of water. Survival time is greatly reduced in hot climates when perspiration losses are at least ½ litre per hour.

Excess water is eliminated—as dilute (pale) urine—by the kidneys. Water in food and drink is severely restricted in kidney failure, but drinking plenty of water is otherwise harmless, and may benefit some people. Some types of kid-

ney stones (see Calcium) are less likely to form if urine is
kept dilute by drinking at least 2 litres of water daily, and
water with meals helps satiety in slimming diets. Water intoxication (really **sodium** depletion) is only likely when
severe perspiration losses are replaced with salt-free beverages.

WHEAT—see also Flour, Gluten, Bread

Wheat is a valuable staple food, providing **energy**, **protein**,
B vitamins and **minerals**. In the average British diet, it supplies a quarter of the total energy and nearly a quarter of the
total protein. Like other **cereals** however, it cannot support
life when eaten alone because it is deficient in vitamins A,
C and B_{12}.

The wholegrain is made into breakfast cereals, and was
formerly used to make a pudding called frumenty. Most
breakfast cereals contain negligible quantities of B vitamins
—they are destroyed during flaking, puffing and toasting.
Some manufacturers of brand name products add three B
vitamins (thiamin, riboflavin and nicotinic acid), when they
are declared on the label. Wheat cereals are a better source
of fibre than other breakfast cereals. See also Bran.

Wheat germ, recovered when white flour is milled, is a
rich source of B complex vitamins and vitamin E. It has a
high unsaturated **fatty acid** content and is liable to become
rancid unless kept under refrigeration. Processed germ
(Bemax) is treated to prevent rancidity: virtually all nutrients
are retained.

WINE AND CIDER

Although wine is strictly prepared from grapes, a variety of
other vegetables and fruits are used to make country wines
and cider. All have a similar nutritional value: they contain **alcohol**, sugars (both are sources of **energy**), and small
quantities of B vitamins and minerals.

Average nutrients in 25 grams (1 oz) of some wheat products

Nutrient	Whole wheat (English)	Puffed wheat	Shredded wheat	Flaked wheat	Wheat biscuits	Wheat germ
Energy (kilojoules)	350	375	380	395	370	325
(kilocalories)	85	90	90	95	90	80
Protein (grams)	2¼	3½	2½	2½	2¾	6½
Carbohydrate (grams)	18¼	19	19¾	19¾	19¼	9
Fat (grams)	½	½	¾	½	½	2
Sodium (milligrams)	1	1	4	—	80[a]	2
Potassium (milligrams)	90	105	75	—	85	290
Calcium (milligrams)	10	10	10	5	10	15
Magnesium (milligrams)	25	30	30	—	30	70
Iron (milligrams)	0·8	0·8	1·1	1	1	1·7
Phosphorus (milligrams) (1)	85	80	70	60	70	230
Other minerals	Small amounts					
Vitamins A, C, D	0	0	0	0	0	0
Vitamin E (α tocopherol) (milligrams)	0·25	0·1	0·1	0·1	—	5·5
B vitamins:						
thiamin (milligrams)	0·1	Trace	0·05[b]	Trace	Trace[b]	0·35
riboflavin (milligrams)	0·04	0·01	0·03[b]	0·01	0·02[b]	0·14
nicotinic acid (milligram equivalents)	0·45	0·7	0·5[b]	0·5	0·5[b]	5·4
pyridoxine (milligrams)	0·1	—	—	—	—	0·24
pantothenic acid (milligrams)	0·4	—	—	—	—	—
biotin (micrograms)	1·3	—	—	—	—	—
folic acid (micrograms)	—	—	—	—	—	—
vitamin B_{12}	0	0	0	0	0	0

– salt added in processing
b – if enriched, thiamin 0·17 milligrams, riboflavin 0·25 milligrams, nicotinic acid 3 milligram equivalents
— – no figures available
(1) At least half the phosphorus is as phytic acid which interferes with the absorption of iron, magnesium and calcium

All table wines contain less than 12 grams of alcohol per 100 millilitres (a small glass): fermentation stops when this level is reached. There is some sugar left in sweet wine, but very dry wines are virtually **carbohydrate** free. Wines contain many other substances in small amounts—for instance, acids, tannins, higher alcohols (congeners), gums, pectin, flavouring oils, glycerol, and esters (combinations of acids and alcohols)—which are responsible for colour, flavour and body. Sulphur dioxide is a permitted **preservative**, in wine.

Sherry, port, madeira, and marsala, fortified with spirit to preserve them, contain up to 17 grams of alcohol per 100 millilitres. Sherry is basically a very dry wine (it is fortified after fermentation is complete) and some sherries are almost carbohydrate free. Most however are later blended with sugar. Vermouths—wines flavoured with bitter ingredients and sugar—are similar in alcoholic strength to the fortified wines.

Cider is less alcoholic than wine. Most contain 4 grams of alcohol per 100 millilitres, though vintage ciders may have twice this amount. Most bottled ciders contain between $2\frac{1}{2}$ to 4 grams of carbohydrate per 100 millilitres, but others vary from 1 to 8 grams per 100 millilitres. Some ciders are as acid as young wine: though lacking in tartaric acid (the chief acid in grapes) they contain malic acid.

Red wine, sometimes thought to be good for blood formation, does not necessarily contain more **iron** than white. It is uncertain how much of the iron in wine is absorbed in normal circumstances, but large intakes (2 litres or more), coupled with a low protein diet, may cause excessive absorption and result in liver damage.

YEAST

Yeast contains **protein**, B **vitamins** and **minerals**. Various types, with differing nutrition properties, are propagated for animal feeds, brewing, baking, **extract** manufacture and

AVERAGE NUTRIENTS IN 10 GRAMS DRIED YEASTS

Nutrient	Bakers	Brewers	Torula
Energy (kilojoules)	70	90	85
(kilocalories)	15	20	20
Protein (grams)	4	5	4·5
Carbohydrate, fat	Trace	Trace	Trace
Iron (milligrams)	1·5	2	2
Potassium (milligrams)	200	170	200
Sodium (milligrams)	5	10	1
Magnesium (milligrams)	20	15	25
Calcium (milligrams)	10	10	10
Vitamins A, C	0	0	0
Phosphorus (milligrams)	15	20	15
Other minerals	—	—	—
B vitamins:			
thiamin (milligrams)	0·15	1	0·2–1·5
riboflavin (milligrams)	0·6	0·4	0·6
nicotinic acid			
(milligram equivalents)	9·7	4·3	4·4
pyridoxine (milligrams)	0·3	0·3	0·3
pantothenic acid (milligrams)	1·5	1	1
others	—	—	—

— – no figures available

vitamin D_2 production. It is also a possible source of novel proteins.

Though used to supplement B vitamin contents of diets, yeast should be avoided by people with a tendency to gout— it is high in purines. Live (active) yeast cannot be eaten in great amounts—it is a laxative and can multiply in the digestive system, sequestering B vitamins in other food— but all nutrients are available from dried (powdered) yeast. Average nutrient contents of dried yeasts are shown in the table above. Dried brewers' yeast supplies more **thiamin** than bakers': about 10 grams will meet the adult **recommended intake** but more is needed to satisfy intakes for other B vitamins. Torula (food) yeasts vary in B vitamin content, depending on the species used.

AVERAGE NUTRIENTS IN SOME WINES

Nutrient	In 150 mls (1 glass) table wine				In 300 mls (½ pint) cider			In 50 mls (small glass) fortified wine		
	Red Medoc	Beaujolais	White Graves	Sauternes	Dry	Sweet	Vintage	Dry sherry	Sweet sherry	Port
Energy (kilojoules)	395	430	460	585	465	530	1260	240	285	320
(kilocalories)	95	100	110	140	110	125	300	55	70	75
Protein (grams)	¼	¼	¼	¼	Trace	Trace	Trace	Trace	¼	Trace
Alcohol (grams)	13	14	13	15	11	11	32	8	8	8
Carbohydrate (grams)	¼	¼	5	9	8	13	22	¼	3½	6
Fat (grams)	0	0	0	0	0	0	0	0	0	0
Sodium (milligrams)	10	15	30	20	20	20	5	110	0	2
Potassium (milligrams)	160	165	130	165	215	215	290		85	50
Iron (milligrams)	1·9	1·1	1·8	0·9	1·5	1·5	0·9	0·2	0·2	0·2
Magnesium (milligrams)	15	15	15	15	10	10	10	5	5	5
Calcium (milligrams)	10	10	20	20	25	25	15	5	5	5
Phosphorus (milligrams)	20	15	10	20	10	10	30	5	5	5
Other minerals	Small amounts	0	0	0	0	0	0			
Vitamins A, C, D	0	0	0	0	0	0	0	0	0	0
B vitamins:										
thiamin (milligrams)	0·008	0·008	0·005	0·005	Trace	Trace	Trace	0·002	0·002	0·002
riboflavin (milligrams)	0·03	0·03	0·02	0·02	0·06	0·06	0·06	0·005	0·005	0·005
nicotinic acid (milligram equivalents)	0·2	0·2	0·1	0·1	0·2	0·2	0·2	0·04	0·04	0·04
pyridoxine (milligrams)	0·08	0·08	0·03	0·03	0·02	0·02	0·02	—	—	—
pantothenic acid (milligrams)	0·06	0·06	0·05	0·05	0·3	0·3	0·3	—	—	—
biotin (micrograms)	—	—	—	—	1·2	1·2	1·2	—	—	—
Other vitamins	—	—	—	—	—	—	—	—	—	—

— — no figures available

YOGHURT

Natural whole milk yoghurt has a similar nutritional value to whole boiled **milk**. Low fat and fat free yoghurts are made from skim milk powder: they have slightly higher **carbohydrate** and protein contents compared with whole milk yoghurts. Whole fruit yoghurts have added sugar—see table below—as well as fruit.

AVERAGE CONTENTS PER 5 OZ (150 GRAM) CARTON

Type	Protein grams	Fat grams	Carbohydrate grams	Energy kilo-joules	kilo-calories	Minerals and Vitamins
Natural whole milk	4·5	5·5	6	400	95	
Natural low fat	5·5	2	8·5	315	75	As for
Natural fat free	6	0·5	10·5	295	70	boiled
Whole fruit low fat	5	2	22·5	525	125	milk

Yoghurt is made by adding a culture of special bacteria to boiled milk which is kept warm for several hours. The bacteria multiply and convert milk sugar (lactose) into lactic acid, which suppresses the growth of harmful bacteria and partially curdles the milk. The milk is preserved (it keeps for about 2 weeks under refrigeration) and thickened. Originally the culture was naturally present in the milk and varied in different areas: some other cultures (for instance in kumiss) contain yeasts which produce alcohol.

Commercial yoghurts contain two bacteria (*Lactobacillus bulgaricus* and *Streptococcus thermophilus*) which complement each other's growth. Most bought yoghurt can be used for home-made yoghurt, but fresh pasteurised milk must be boiled before incubating it, otherwise harmful bacteria may become established before sufficient acid is formed to inhibit their multiplication. Dried skim milk can be added for a firmer, less acid, curd. All (except pasteurised) commercial yoghurts contain living lactic acid bacteria, but so called 'live' ones may contain others which can spoil the flavour.

About 70 years ago, Metchnikoff claimed that life could be prolonged by eating plenty of yoghurt. The lactic acid

bacteria would inhibit the growth of putrefactive bacteria—
which were supposed to form harmful toxins—in the large
bowel. His claims were refuted, but yoghurt has not lost its
reputation as an especially 'healthy' food. It is as nutritious
as milk. Any diet containing much lactose will promote the
growth of lactic acid bacteria in the digestive system.

Acidophilus milk is also yoghurt, but *L. bulgaricus* is
placed with *Lactobacillus acidophilus*, which normally
inhabits the large bowel in humans. Acidophilus milk (or
a pharmaceutical preparation of drug resistant *L. acidophilus*)
may help to check the diarrhoea which sometimes follows
sterilisation of the intestine with antibiotics.

ZINC

A **trace element** necessary for growth, sexual maturity,
wound healing, and taste and flavour perception.

Approximate zinc contents of some foods are shown in
the table overleaf. Animal foods—milk, meat and fish—are
the best sources, and **amino acids** in animal proteins may
enhance its absorption in the blood stream. It appears to be
less well digested from plant foods and vegetarians may
require more zinc than meat eaters. Wholegrain cereals
and pulses are apparently good sources, but they contain
phytic acid which probably interferes with absorption. More
zinc is thought to be available from wholemeal bread
leavened with yeast, which partially degrades phytic acid
during proving. White flour is a poor source, but is also low
in phytic acid. Vegetables and fruits are also poor.

Most adults probably require about 10 milligrams of zinc
daily—an amount contained in average British diets—but
there is too little available information to set reliable recommended intakes. Adolescents and pregnant and breast feeding mothers have the highest requirements—the increased
needs should be met with extra meat, milk and fish.

Less than 40% of zinc in the diet is thought to be transferred to the blood stream during **digestion**. Only 10% may

be absorbed from strict vegetarian diets. Adults contain 2 to 3 grams of zinc, but most is retained in the skeleton and there appear to be no large stores that can be drawn upon when zinc is in short supply. The remainder is inside cells,

SOME SOURCES OF ZINC (1)

Rich sources	more than 5 milligrams zinc in 100 grams food: Oysters, wheat germ, wheat bran, cocoa
Good sources	3 to 5 milligrams zinc in 100 grams food: Beef, lamb, liver, cheese, dried skim milk, peanuts, oatmeal
Moderate sources	1 to 3 milligrams zinc in 100 grams food: Pork, veal, eggs, turkey, tuna, wholegrain cereals (like wholemeal bread), dried pulses
Poor sources	less than 1 milligram zinc in 100 grams food: Most fruit, most vegetables, cooking fats, oils, butter, sugar, refined cereals (like cornflakes, white bread), milk, white fish, chicken breast, salmon

(1) based on figures given by Murphy, Wells and Watt (1975) *J. Am. Diet. Ass.* **66** p. 345

where it is essential for the action of many **enzymes**—including those concerned with the formation of new body **proteins** for growth and repair, and with cell division and multiplication. Extra zinc is therefore needed for growth and wound healing. The young born to female animals fed a diet lacking in zinc during pregnancy are deformed—a finding that may be of importance to humans.

Zinc is lost from the body in urine and sweat. In parts of the Middle East adolescents are known to suffer from dwarfism and sexual immaturity which can be prevented by improvement in diet or zinc supplements. These children were eating very restricted diets—mainly wholemeal unleavened bread—and, in the hot climate, were losing excessive amounts of zinc in sweat. Deficiency is made worse by loss of blood (from hookworm infection) and probably too little protein. Such severe deficiency is unlikely in Britain, but mild deficiency, affecting taste and smell—which would impair appetite and growth, has been detected in some American children.

Malabsorption can also cause deficiency, and there are large losses from the body in alcoholism, artificial kidney

machine treatment, fever, and after surgery or burns. Unless
extra zinc is given recovery may be delayed by lack of appetite and wounds may not heal. In animals, excess **copper** or
cadmium (see Poisons in food) can displace zinc in the
body, but it is not known if deficiency can occur in this way
in man.

Supplements of zinc are inadvisable (except for the treatment of deficiency): in animals excess zinc can cause copper
deficiency, and interfere with the utilisation of iron, causing
anaemia. Very large intakes are toxic: foods and water contaminated with zinc from galvanised metal containers has
caused outbreaks of vomiting.

APPENDIX

RECOMMENDED DAILY INTAKES OF ENERGY AND NUTRIENTS FOR THE UK (1)

Age	Energy kilo-calories	Energy mega-joules	Protein grams	B vitamins Thiamin milligrams	B vitamins Riboflavin milligrams	B vitamins Nicotinic acid milligram equivalents	Vitamin C milligrams	Vitamin A microgram equivalents	Vitamin D micrograms	Calcium milligrams	Iron milligrams
Boys and Girls											
Up to 12 months	800	3·3	20	0·3	0·4	5	15	450	10	600	6
1 year	1200	5·0	30	0·5	0·6	7	20	300	10	500	7
2 years	1400	5·9	35	0·6	0·7	8	20	300	10	500	7
3 to 4 years	1600	6·7	40	0·6	0·8	9	20	300	10	500	8
5 to 6 years	1800	7·5	45	0·7	0·9	10	20	300	2·5	500	8
7 to 8 years	2100	8·8	53	0·8	1·0	11	20	400	2·5	500	10
Boys											
9 to 11 years	2500	10·5	63	1·0	1·2	14	25	575	2·5	700	13
12 to 14 years	2800	11·7	70	1·1	1·4	16	25	725	2·5	700	14
15 to 17 years	3000	12·6	75	1·2	1·7	19	30	750	2·5	600	15
Girls											
9 to 11 years	2300	9·6	58	0·9	1·2	13	25	575	2·5	700	13
12 to 14 years	2300	9·6	58	0·9	1·4	16	25	725	2·5	700	14
15 to 17 years	2300	9·6	58	0·9	1·4	16	30	750	2·5	600	15

| Age | Energy | | Protein grams | B vitamins | | | Vitamin C milligrams | Vitamin A microgram equivalents | Vitamin D micrograms | Calcium milligrams | Iron milligrams |
	kilo-calories	mega-joules		Thiamin milligrams	Ribo-flavin milligrams	Nicoti-nic acid milligram equivalents					
Men											
18 to 34 years Sedentary	2700	11·3	68	1·1	1·7	18	30	750	2·5	500	10
Moderately active	3000	12·6	75	1·2	1·7	18	30	750	2·5	500	10
35 to 64 years Very active	3600	15·1	90	1·4	1·7	18	30	750	2·5	500	10
Sedentary	2600	10·9	65	1·0	1·7	18	30	750	2·5	500	10
Moderately active	2900	12·1	73	1·2	1·8	18	30	750	2·5	500	10
65 to 74 years Very active	3600	15·1	90	1·4	1·7	18	30	750	2·5	500	10
Sedentary life	2350	9·8	59	0·9	1·7	18	30	750	2·5	500	10
75 and over Sedentary life	2100	8·8	53	0·8	1·7	18	30	750	2·5	500	10
Women											
18 to 54 years Most occupations	2200	9·2	55	0·9	1·3	15	30	750	2·5	500	12
55 to 74 years Sedentary life	2050	8·6	51	0·8	1·3	15	30	750	2·5	500	10
75 and over Sedentary life	1900	8·0	48	0·7	1·3	15	30	750	2·5	500	10
Pregnancy, 3 to 9 months	2400	10·0	60	1·0	1·6	18	60	750	10a	1200b	15
Breast feeding	2700	11·3	68	1·1	1·8	21	60	1200	10	1200	15

a – for all 9 months
b – last 3 months
(1) Based on Recommended Intakes of Nutrients for the United Kingdom DHSS Rep. Pub. Health & Med. Sub. no. 120 1969

BIBLIOGRAPHY

Agricultural Research Council/Medical Research Council (1974), *Food and Nutrition Research*. London: HMSO.

Amor (1975), *A Guide to Food Regulations in the UK*. Leatherhead: British Food Manufacturing Industries Research Association.

Bender (1960), *Dictionary of Nutrition and Food Technology*. London: Butterworths. BNF Bulletins 1–15, London: British Nutrition Foundation.

Committee on Food Protection (1966), *Toxicants Occurring Naturally in Foods*, Food and Nutrition Board, National Research Council. Washington DC: National Academy of Sciences.

Davidson, Passmore, Brock and Truswell (1975), *Human Nutrition and Dietetics*. Edinburgh: Churchill Livingstone.

Department of Health and Social Security (1969). Rep. Pub. Health Subj. 120, *Recommended Intakes of Nutrients for the United Kingdom*. London: HMSO.

Department of Health and Social Security (1969). Rep. Pub. Health and Med. Subj. 122, *The Fluoridation Studies in the UK and the Results Achieved after Eleven Years*. London: HMSO.

Department of Health and Social Security (1974). Rep. on Health and Social Subj. 7, *Diet and Coronary Heart Disease*. London: HMSO.

Diem and Lentner (Eds) (1970). *Scientific Tables*. Basle, CIBA-GEIGY Ltd.

Drummond and Wilbraham (1939), revised Hollingsworth (1958), *The Englishman's Food*. London: Cape.

Food and Agricultural Organisation (1970), *Amino Acid Contents of Foods and Biological Data on Proteins,* Nutritional Studies No. 24. Rome: FAO.

Forbes (Ed.) (1971), *Joint Survey of Pesticide Residues in Foodstuffs sold in England and Wales, 1967–8.* London: Association of Public Analysts.

Goodwin (Ed.) (1967), *Chemical Additives in Food.* London: Churchill.

Harris and von Loesecke (1960), *Nutritional Evaluation of Food Processing.* Chichester: J. Wiley and Sons Inc.

Hollingsworth and Russell (Eds) (1974), *Nutritional Problems in a Changing World.* Essex: Applied Science.

Hilditch and Williams (1964), *The Chemical Composition of Natural Fats.* London: Chapman and Hall.

Marks (1968), *The Vitamins in Health and Disease.* London: Churchill.

Merory (1968), *Food Flavourings, Composition, Manufacture and Use.* Westport: Avi Publishing Inc.

Lucas (1975), *Our Polluted Food, a Survey of the Risks.* London: Charles Knight.

Ministry of Agriculture, Fisheries and Food, (1966) Food Standards Committee, *Report on Claims and Misleading Descriptions.* London: HMSO.

Ministry of Agriculture, Fisheries and Food (1970), *Manual of Nutrition.* London: HMSO.

Ministry of Agriculture, Fisheries and Food, Working Party on the Monitoring of Foodstuffs for Heavy Metals. London: HMSO:

> First Report (1971) *Mercury in Food*
> Second Report (1972) *Lead in Food*
> Third Report (1973) *Mercury in Food*
> Fourth Report (1973) *Cadmium in Food.*

Ministry of Agriculture, Fisheries and Food (1974), Food Standards Committee, *Second Report on Bread and Flour.* London: HMSO.

Ministry of Agriculture, Fisheries and Food (1974), Food

Standards Committee, *Report on Novel Protein Foods*. London: HMSO.

O'Keefe, *Bell's Sale of Food and Drugs* (14th Ed.). London: Butterworth and Co.

Platt (1962), *Tables of Representative Values of Foods Commonly used in Tropical Countries*, Medical Research Council Spec. Rep. Ser. 302. London: HMSO.

Pearson, *The Chemical Analysis of Foods* (6th Ed.). London: Churchill.

Pyke (1964), *Food Science and Technology*, London: Murray.

Sinclair and Hollingsworth, *Hutchinson's Food and the Principles of Nutrition* (12th Ed.). London: Arnold.

Stein (Ed.) (1971), *Vitamins, a University of Nottingham Seminar*. London: Churchill Livingstone.

West, Todd, Mason, Van Bruggen (1966), *Textbook of Biochemistry*. New York: Macmillan.

World Health Organisation Tech. Rep. Ser. (1973) No. 522, *Energy and Protein Requirements*. Geneva: WHO.

World Health Organisation Tech. Rep. Ser. (1973) No. 532, *Trace Elements in Human Nutrition*. Geneva: WHO.

Reference has also been made to articles in various journals, including:

The American Journal of Clinical Nutrition; The British Journal of Hospital Medicine; Journal of the American Dietetic Association; British Journal of Nutrition; New Scientist; The Lancet; The British Medical Journal; Nutrition; Nutrition and Food Science; Nutrition Reviews; The Practitioner; The Proceedings of the Nutrition Society and The World Review of Nutrition and Dietetics.